GRAND UNIFIED THEORY

GRAND UNIFIED THEORY

THE GREATEST DISCOVERY OF THE 21ST CENTURY OPENING THE DOORS TO A NEW WORLD

ARKAY NAIR

TATE PUBLISHING
AND ENTERPRISES, LLC

Published by Tate Publishing & Enterprises, LLC
127 E. Trade Center Terrace | Mustang, Oklahoma 73064 USA
1.888.361.9473 | www.tatepublishing.com

Tate Publishing is committed to excellence in the publishing industry. The company reflects the philosophy established by the founders, based on Psalm 68:11,
"The Lord gave the word and great was the company of those who published it."

Cover design by Joseph Emnace
Interior design by Gram Telen

Published in the United States of America

ISBN: 978-1-63122-827-8
1. Science / Physics / Quantum Theory
2. Science / Mechanics / General
14.06.11

DEDICATION

To the greatest mind in physics and mathematics—
Sir Isaac Newton

CONTENTS

BOOK I

MY DISCOVERY OF THE GRAND UNIFIED THEORY

INTRODUCTION

I have been indeed fortunate to realize the Grand Unified Theory. This blissful event happened to me after forty years of contemplation on this subject in my spare time since my graduation from college with a Bachelor's degree in Science with first rank in the university. Albert Einstein was able to discover the mathematical formula $E=mc^2$, but his explanations and inferences from his discovery are not complete. My humble endeavor here is to complete his work and present this paper to you for your review. As a result of this discovery, we are now in a position to solve many of the puzzles and paradoxes that have been challenging scientists for the last four hundred years. With this discovery, we are now in a position to solve many of our most urgent problems of the day, like global warming and energy shortages.

Book I introduces the reader to a unified theory in the form of a research paper consisting of five discourses. Each discourse focuses on one major aspect of this theory. The first two discourses introduce ideas that are contrary to current popular beliefs. I request that you bear with me and read through these sections without pausing to formulate questions. The third and fourth discourses elaborate on these controversial ideas in formulating a unified theory. The fifth discourse introduces a definition of the unified theory from my viewpoint; and invites readers to formulate their own view of this theory. It is essential that these five discourses be read in chronological order, from the first to the fifth, and that they are read in one session without interruption. Including this introduction, Book I is 6,769 words long. Therefore, please start reading this section only when you have sufficient uninterrupted time to complete the reading without break. Please hold all your questions until a second reading. Many of your questions are answered along the way as you read and may become irrelevant by the time you finish reading. In case you do have genuine questions after a complete reading, please send them to my email address at rass.nair@yahoo.com.

Each discourse begins with an abstract where the central idea is introduced. This is followed by a discourse where the idea is analyzed and discussed under the assumption that the reader is already familiar with the current popular views on the subject. The discourse ends with a summary where the idea presented is reinforced; and establishes the basis of arguments for the succeeding discourses.

Let me present you with an unconventional idea: please read Book I as if it is written by you. The expression "My Discourse" applies to the writer when he is writing it. Now that I have finished writing and am presenting my writings for your review, this paper really belongs to the reader when he or she is reading it. I am sure that my very thoughts have gone through your mind at one time or the other. My job here is to present these ideas to you logically so that you can see the real picture and separate facts from fiction. If you are ready to read *your* Grand Unified Theory, you may begin now!

MY DISCOURSE ON GEOCENTRIC UNIVERSE

ABSTRACT

The earth is the center of our universe. Ptolemy is right; Tycho Brahe is right. Copernicus is wrong. Galileo and Kepler perpetuated the mistake that Copernicus made. Mankind has been moving in the wrong direction for the last four hundred years because Copernicus failed to observe an important principle of scientific pursuit, which we may call by the name "Observer Equivalency." A Thought Experiment is used to explain Observer Equivalency in identifying four types of observers: Earth, Earthling, Alien, and Sky. Conclusive experimental proof for a geocentric universe comes from the famous Michelson–Morley experiments.

DISCOURSE

Mankind had believed in a geocentric universe for a long time. In the second century AD, Ptolemy recorded his observations in the *Almagest*, which relied heavily on the works of earlier Greek astronomers and was dependable for all practical purposes. However, fourteen hundred years later, Copernicus challenged this theory by reintroducing an old, discarded theory of a heliocentric universe mainly based on two problems—the retrograde motions of the outer planets and the complexity of calculating the epicycles.

Tycho Brahe used improved instruments and devoted his full time for observations and calculations. He suggested an improvement on the Ptolemaic system by proposing that the moon and the sun revolved around the earth, which is fixed at the center, but all other planets revolved around the sun. This theory is fully supported by observations; and explains all heavenly motions, including the elliptical revolutions of the other planets around the sun as it circles around the fixed Earth at the center. Copernicus's heliocentric theory is purely a mathematical model that does not agree with observed phenomena. His theory is only applicable to a solarian (an inhabitant of the sun), if there is such a creature! Copernicus, Kepler, and Galileo failed to define the observer and the place of observation when making these observations. Depending on who is the observer and from where he or she is observing, observed results are going to be different. When our astronauts visited the moon, they saw the earth orbiting the moon, based on the principle of observer equivalency.

We must conduct a thought experiment to understand observer equivalency. Let us think about a common housefly with two large compound eyes on its head. Let us fuse these two eyes into one large eye covering half of its head. Let us now make it large enough to cover its whole head. Finally, let us make the eye still larger so that its entire body is one spherical eye covering its whole body surface. This is the model of the earth as an observer. Imagine the whole surface of the earth as a spherical compound eye with 360-degree vision towards the sky and the center of the earth as its brain. What is the inference of this observer about the heavens? *The orbital motions of the heavenly bodies are not observed by the earth.*

In other words, the earth does not experience the dawn and the dusk that an earthling observes. Instead, it is always high noon for the earth! An Earthling, on the other hand, can see about two radian angles (less than 120 degrees) at a time with the best of peripheral vision; and he occupies a single point space on the surface of the earth resembling a single eye unit of Earth's compound eye. Unlike Earth, an Earthling *does* experience a dawn and dusk, and no two earthlings would agree about their independent observational results. A Tokyo observer would insist that he is witnessing dusk (setting of the sun) while at the same time a Boston observer would swear that he is witnessing dawn (rising of the sun). The earth would always see the sun directly overhead. Earth as an observer is different from an

Earthling as an observer. Earth's vision is the sum total of all individual Earthlings' viewpoints combined.

An alien observer is defined as an observer who is not Earth-based. What an alien observer sees is different from what an Earthling sees or the Earth sees as a whole. Since we have assigned the Earth the central position, the alien is always positioned at the periphery. A solar observer (solarian) would be making a true statement when he says that the sun is the center of *his* universe. So would be the statement of a Martian's or a Venusian's. The sky similarly may be imagined to have a compound eye, with a vision from all points in the sky, looking down on Earth from all points inside the celestial sphere. The vision of the sky would be the sum total of all individual alien's viewpoints combined.

The observer is an integral part of observation and inference. As a major in chemistry, I am quite familiar with the methodology that chemists follow. We use the terminology "experiment, observation, and inference." A chemist records his inferences based on his observations and not those of anyone else. Physicists do not seem to care much to adhere to this principle, and therefore their writings are prone to become fiction. Copernicus is the first science fiction writer, who has been feeding our brains with fiction for the past four hundred years.

Apart from the fact that the observer is an Earthling, we must also realize that we are not aware of the existence of any other observer in our universe. No life form, let alone intelligent life, has been found in any other planet or satellite. Therefore, it is perfectly logical to infer that Earth is the center of our universe because life is likely to form only at the center of any universe. Stated differently, the observer tends to migrate to the center of his or her field of observation. More importantly, with the discovery of more matter within the solar system since Copernican times, earth qualifies as the *barycenter* of the solar system (our universe). This physical model is therefore superior to the heliocentric model. Copernican heliocentric model is only justified mathematically. It is not a physical possibility.

The Michelson-Morley experiment provides conclusive proof that Earth is fixed in space, and therefore the speed of light is the same in all directions on the surface of the earth. The scientific community, including the top scientist of the day, Einstein, failed to reach the correct inference from this experiment—which is that Earth does not rotate or revolve. Mathematically speaking, Earth's rotation and revolution are to be assigned a value of zero. Being a real number, it is perfectly logical to assign this value to Earth's revolutional and rotational speeds. Now, we face a tremendous task of recalculating all physical measurements based on the heliocentric theory that we have been using for the last four hundred years to be converted to a geocentric model which is the true physical reality.

The Michelson-Morley experiment also proved that the speed of light is a constant. This discovery resolved the problem of explaining the diurnal revolution of the distant stars around a tiny Earth simply by adding two words to the statement of this phenomenon as follows: *Lights from* the distant stars make a complete revolution around Earth once in every twenty-four hours.

SUMMARY

The earth is the center of "our" universe. For the first sixteen hundred years of the Christian era, we believed so and used that knowledge in our everyday lives and activities. We then decided to rebel against this view and embraced a heliocentric view, even though we did not have any empirical proof to change our view. Therefore, it is logical and scientific to go back to a geocentric system and recalculate our data and correct our mistakes of the past four hundred years. The task is enormous, but it is achievable. The rewards are going to be enormous. We can solve many of our major problems, including global warming and energy shortages, once we attain the correct perspective about our orientation as central in our universe. The terminology "Planet Earth" should be changed. We must call our home "Center Earth" or some other suitable name. The concept of Center Earth is necessary to understand the Grand Unified Theory. The reader should therefore now assume the mindset that Earth is the center of our universe in order to understand the following discourses.

READER'S NOTES

MY DISCOURSE ON THE STRUCTURE AND FUNCTION OF THE UNIVERSE

ABSTRACT

The universe is structured like Niels Bohr's model of the atom. Earth corresponds to the nucleus. The moon constitutes the primary shell, which may be called the "lunar shell" or "lunar orbital." The sun and its planets, excluding Earth, form the next shell or orbital, which may be called the "solar shell" or "solar orbital." Next is the "galactic shell" or the "galactic orbital," which constitutes all heavenly bodies beyond the solar system. The earth as the center is assigned a rotational and revolutionary speed of zero; meaning that it is fixed and does not revolve or rotate. Matter is defined as "that which occupies space." Space, in turn, is defined as "that which accommodates matter." There is no such thing as vacuum or empty space. The elusive aether is nothing but hydrogen in plasma state. With regard to their relationship, matter and energy are defined as follows: Matter is energy at rest, energy is matter in motion. All forms of electromagnetic radiation are plasma hydrogen in motion at speeds up to the limit of "c." The double-slit experiment gives conclusive proof of the existence of aether as plasma hydrogen.

DISCOURSE

Atoms are invisible. Even the largest among them, uranium, cannot be seen by the naked eye or even with the most powerful microscopes. Niels Bohr proposed a model of the atom based on the model of the universe as it is observed holistically. In this planetary model, when we use our geocentric model proposed in the first discourse, Earth corresponds to the nucleus of the atom. The moon and the sun constitute the first and second shells, or orbitals, respectively called the lunar shell and the solar shell. The stars and galaxies beyond the solar system are all included in one shell called the galactic shell or galactic orbital. Niels Bohr proposed a planetary model for the atom because he felt that the building blocks of the universe should look like miniature units of the same. In other words, atoms are miniature models of the universe, and the universe in turn looks like a giant atom. Just as a wall made up of rectangular bricks would look like a huge rectangular brick, a universe made up of atoms should look like a huge atom.

In this physical model, the sun wobbles around a stationary Earth once in twenty-four hours. Wobbling motions are characteristic of stars, and the sun does the same around its barycenter. The moon circles Earth in about twenty-four hours and fifty minutes. All other planets and their moons circle the sun at varying speeds and distances, observing Newton's laws of motion and universal law of gravitation. Light, which is corpuscular in nature as Newton proclaimed, is actually matter in plasma form (mainly hydrogen). Light obeys gravitational law and falls onto the most massive body within the solar system, the sun, preferentially. This inference explains why stars are not visible during daytime and why they are visible at night, casting a stream of light along the path of the sun.

In considering the relationship between matter and space, we see that the following definitions fit well: Matter is defined as that which occupies space. Space is defined as that which accommodates matter. The function of space is to accommodate matter; matter in turn occupies all available space. This means that the term "empty space" is a misnomer. There is no such thing as empty space. Interstellar and intergalactic space is filled with matter in its thinnest form—plasma hydrogen. We know that interstellar space has a temperature of a little under 3 degrees Kelvin. Applying Boyle's Law of Gases and the Ideal Gas Law ($PV = nRT$), we can easily see that interstellar space V (volume) is enormous, since P (pressure) and T (temperature) are very low values and R (the gas constant) is a constant. This volume is occupied by an extremely thin form of matter at an extremely low pressure. This realization solves the mystery of the missing dark matter that we need in order to prevent the universe from expanding forever! This realization also proves that there is, after all, such a thing as aether, which is the fourth state of matter, "plasma." A major component of this plasma matter is hydrogen, but it may contain plasma forms of other elements as well. Einstein's photons are nothing but plasma hydrogen or plasma helium moving at speed c or lower. Therefore, Newton's corpuscular theory of light is still valid. Light behaves as matter *and* wave; not as matter *or* wave.

In considering the relationship between matter and energy, the following definitions apply*:* Energy is matter in motion, and matter is energy at rest. This hypothesis is consistent with our observation. We see the earth at rest, and we see the heavenly bodies in motion. Our knowledge of the heavenly bodies comes to us in the form of energy: light, heat, electricity, and magnetism. Stated differently, all heavenly bodies touch the earth through light, heat, or electromagnetic waves. A better way to visualize this is to see the world as a huge ocean of plasma hydrogen with all the bodies suspended in it. This would constitute a holistic model of the universe.

In describing the structure of our geocentric universe, let us assign the earth (the nucleus) a fixed status with rotational and revolutionary speeds of zero. Zero is a real number. Therefore, we are mathematically correct in assigning this value for Earth's rotation and revolution just as we do in algebraic problem solving. We must now recalculate all other values presently in use under the premise of a heliocentric universe. All calculations, especially those that use parallax measures based on an assumed revolution of the earth around the sun; have to be recalculated.

Thomas Young's original double-slit experiment, along with all similar experiments conducted later under varying conditions, have been a puzzle to all scientists who were tutored in classical mechanics. If you accept the existence of aether (plasma) occupying all space, it is easy to make the correct inference that light creates waves in plasma medium. Stated differently, light, and all other electromagnetic waves are ripples in plasma medium. The best way to understand this phenomenon is to compare light to water. Water makes the ocean; water also makes the ocean waves. You don't see any ocean waves in the Sahara desert because there is no ocean there. Light also behaves the same way. There is no reason to believe that light behaves differently from water or any other matter. Sound waves need air for their propagation. Light, whether we call it by the name of plasma or photon, is matter. We may even call light super-matter since only light can reach speeds up to c (300,000 km/sec.) both as a particle *and* as a wave in its own plasma medium.

Ocean waves need water to propagate. Sound waves need air to propagate. Light needs plasma or aether to propagate. Plasma exists everywhere, even in a vacuum tube. Water flows through a pipe or a fireman's hose when it is in directed motion. When we see waves in the ocean, water molecules are just pushing their neighbors creating the domino effect. When air is in directed motion, we call it wind. When air molecules simply push their neighbors like dominoes, it creates vibrations called sound. When a hydrogen atom (a proton and an electron) is in linear motion, it forms a light beam (a photon, if you prefer). When a hydrogen atom vibrates in a pool of protons and electrons (plasma), we notice the wave-like nature of light. This is the correct inference from the double slit experiment.

SUMMARY

The microworlds and the macroworlds are more likely to be similar than different. Intuition and observation tell us that the universe has to be a macrocosm of the atoms of which it is made of, and vice versa. A geocentric universe may be drawn with the earth as the central nucleus and all heavenly bodies forming lunar, solar, and galactic shells or orbitals. In relation to space, matter is defined as "that which occupies space." In relation to matter, space is defined as "that which accommodates matter." This realization leads us to the inference that space cannot exist without matter in it. Aether does exist, which has to be hydrogen in plasma form. Interstellar space is found to have a temperature of about 2.7 degrees Kelvin. The most logical explanation for this phenomenon is that deep space is not empty. Plasma is the logical choice as the fourth state of matter along with solid, liquid, and gas. Light, heat and all electromagnetic waves travel through space in a medium of plasma or aether. Plasma or aether limits the speed of all electromagnetic waves to a maximum of c by acting as a resisting medium. If we are ready to accept Empedocles' "fire" as a state of matter, we have accounted for all the "missing dark matter" necessary to balance our equations for a stable universe. The double-slit experiment confirms this theory.

READER'S NOTES

MY DISCOURSE ON THE BIG BANG AND BLACK HOLES

ABSTRACT

The Big Bang Theory as the origin of the universe is flawed. Newton's third law of motion states that for every action there is an equal and opposite reaction. If the Big Bang Theory is true, then there should have been a corresponding implosion that created the center of the universe. The principle of observer equivalency is applied to an imaginary observation of the Big Bang by our defined observers, and their inferences are explained from each observer's view point. A modified theory called the "Bing Bang" is proposed to describe the possible formation of the earth in a preexisting infinite universe through a supernova type of explosion/implosion of a companion star of our sun when it reached or exceeded the Chandrasekhar limit.

DISCOURSE

The Big Bang Theory, as it is understood today, is flawed. It does not fit in with the classical laws of physics that we have learned and accepted. Newton's third law of motion states that for every action there is an equal and opposite reaction. If we may use the word "bing" to denote an implosion, we may justifiably ask the question, "Where is the bing?" Did it create the earth? If so, the earth is the center of the universe. Did it create a big black hole in the center of our galaxy or any galaxy? If so, that black hole is the center of the universe. Mathematically speaking, the origin of the universe becomes a closed set if we accept the big bang theory, while the end of the universe remains an open set. Naturally, we may then accept that the universe has a center but no defined periphery. The problem to solve now is whether this center is the Earth or some other material body. Let us apply our logical minds to this question to come up with some possible theories using the principle of observer equivalency that we adopted in a previous discourse.

We, the earthlings, see the universe as finite with the sky as the boundary. Earthlings observe the heavenly bodies, all moving from east to west, uniformly, centripetal to the earth; so that they don't collide with each other and fall to the center. Each body has a free path to travel, similar to the flow of traffic on a highway. The cosmos is also protected by its law of speed limit, c, which no particle is allowed to exceed. According to Hubble, the universe is still expanding because of the observed red-shift of some distant galaxies. Well, what about the blue sky? The blue sky tells us that the universe is contracting! Since we observe both these phenomena, we may safely assume that our universe is in equilibrium, with inflationary tendencies counterbalanced equally with deflationary tendencies. It is therefore safe to assume that the universe is in a steady state of equilibrium. The existence of Cosmic Background Radiation tells us that the sky is falling (not pigeon droppings!), indicating the deflationary tendencies of our universe counterbalancing inflationary tendencies.

Assuming that the Big Bang Theory is true, let us see what our four observers would be seeing if they were fortunate enough to witness this cataclysmic event. The earth would experience the implosive aspect of the Big Bang as it is formed by the crushing force kicked back by escaping plasma matter. The sky and all the aliens living out there would experience the explosive aspect of the Big Bang phenomenon. An earthling, who is right in the middle of this cataclysmic event, would feel the impact of the explosion and the implosion. She will see the earth being formed by the imploding plasma while the heavens are being formed by the exploding plasma. She has an equal chance of being dropped down to the newly formed earth to become an earthling; or of being flung into outer space to become an alien out there.

In my opinion, the Big Bang Theory should only be applied in discussing the origin of the earth and not that of the universe. We do observe supernova explosions and the gradual fading away of a star into oblivion. If we accept the idea of an infinite universe, any search for its origin or demise is futile. However, we can definitely search for the origin of the earth, since we know that it is finite. That is why I propose a hypothesis that the earth was formed in a supernova type of explosion/implosion at some point in time that is calculable (perhaps 4.5 billion years ago). Binary star systems are quite common in the universe, and it is quite possible that our sun had a companion star larger than itself. When that star reached or exceeded its Chandrasekhar limit, it may have exploded, forming the earth as the implosive part of the equation, while an equal amount of matter escaped into the space in the form of electromagnetic radiation.

The earth is now in a state of equilibrium, with explosive forces counterbalanced with equal implosive forces. Hubble's theory of an expanding universe is counterbalanced through the constant shower of cosmic background radiation, which is in fact telling us that the sky is falling! The sky, of course, is made up of nothing other than plasma matter.

SUMMARY

If we believe that the universe is infinite, we cannot assign a beginning to it. If we accept the Big Bang Theory, we must also accept the fact that our universe is finite or semi-finite. A big bang has to be accompanied by a big bing—

by which I mean a corresponding implosion that would form the center, from where matter exploded, according to Newton's third law of motion. It is more logical to search for the origin of the earth, which we can see as finite in an infinite world. The most probable origin of the earth may be through a supernova type of explosion/implosion of a companion star of our sun when it reached or exceeded the Chandrasekhar limit in size.

READER'S NOTES

MY DISCOURSE ON THE FORMATION OF THE ELEMENTS AND COMPOUNDS

ABSTRACT

Earth was formed most likely by a supernova type of explosion of a companion star of our sun. According to Newton's third law of motion, an explosion has to be accompanied by an equal and opposite implosion. As matter flew away into space in the form of plasma, it drew its momentum, by crushing matter left behind, into the center. Plasma hydrogen left behind may first condense into atomic hydrogen. A hydrogen atom subject to further force may form a neutron by the fusion of its proton and electron. Higher elements may successively be then formed with the available pool of protons, electrons, and neutrons. As the pressure is maintained (from the kick of escaping plasma), higher and higher elements are formed with a corresponding decrease in volume. In the case of the earth, this process has continued until ninety-two natural elements were formed. Later, elements entered into chemical reactions among themselves to form compounds.

DISCOURSE

I believe that the earth was formed by a supernova explosion of a companion star of our sun rather than through core accretion. It is then possible that higher elements are formed as a result of the implosive force exerted by escaping plasma. The escaping matter needs a jumping board. In a purely mechanical way, just like the kneading of the dough, plasma hydrogen condenses into atomic hydrogen made up of a proton and an electron. If a proton and an electron are squeezed hard enough, they may fuse to form a neutron. This pool of protons, electrons, and neutrons, the only true possible building blocks of the universe, when subjected to further kneading from the kicks of escaping plasma hydrogen successively condensed into higher elements. At a later stage, elements reacted among themselves to form compounds.

Earth is the densest body in the solar system. The earth, terrestrial in nature, has a mean density close to four times that of the sun, a voluminous body made up mainly of hydrogen and some helium. This indicates the massive implosive force that Earth received in the implosive part of the equation that resulted in the formation of ninety-two natural elements. The last one on this list, uranium, carries as many as 92 protons and over 140 neutrons in its belly protected by a tight shell of 92 electrons.

All chemical reactions tend toward equilibrium. It is therefore quite possible that elements are still undergoing chemical reactions in the bowels of the earth. It is quite possible that carbon and hydrogen are combining to form various hydrocarbon products below the earth's surface as we remove the products of reaction (crude oil and natural gas) through our mining activities. In the earth's biosphere, life, in the form of plants, is doing an excellent job in reversing entropy by converting waste products of water and carbon dioxide into carbohydrates by photosynthesis. I believe that plants give us clues to the mysteries of life. Only green plants with chloroplasts can convert carbon dioxide, which is the waste product of animal respiration with maximum entropy, back into edible food as carbohydrates for all animals and human beings. Only green plants are capable of reversing entropy in the universe. Further research is needed into these areas.

SUMMARY

The most logical hypothesis about the formation of Earth is a supernova type of explosion of a companion star of our sun when it reached or exceeded the Chandrasekhar limit in mass. This explosion had to be accompanied by a corresponding implosion according to Newton's third law of motion. This implosive force is the mechanism by which elements are formed along the periodic table. Elements later combined to form compounds. All chemical reactions tend toward equilibrium. If one of the products of a reaction is removed from the system, the reaction tends to move forward forming more of that product. In the earth's crust, hydrogen and carbon may be combining to form hydrocarbons as we remove the product of the reaction (crude oil) through mining. In the earth's biosphere, life, in the form of plants, seems to be the only force capable of reversing entropy by converting carbon dioxide and water (human and animal waste products from respiration) into carbohydrates (food for animals and human beings). Further research is necessary to take advantage of these concepts to solve our problems in energy, food, and environment.

READER'S NOTES

MY DISCOURSE ON THE GRAND UNIFIED THEORY

ABSTRACT

Based on the adoption of a geocentric universe in three-dimensional space occupied by four states of matter, I have formulated a Grand Unified Theory. The theory is presented in three ways using three sets of terms borrowed from Newton. The reader is encouraged to contemplate on this theory and come up with his or her own version of this theory. This should serve as a personal philosophy of the reader whether he or she is a student, teacher, researcher, or entrepreneur.

DISCOURSE

Science has come a long way in its search for a single theory that would explain all phenomena in nature. Faraday and Maxwell showed us that electricity and magnetism are actually two aspects of the same force we now call electromagnetism. Einstein spent the last three decades of his life searching for a field theory that would unite electromagnetism and gravity. The basic idea of unification is to find some underlying commonality in nature, some property that remains the same while something else is changed.

Twentieth-century scientists missed out on discovering a unified theory because they were searching for it in a fictional world built on heliocentrism. The Copernican heliocentric model is an excellent mathematical model, but nothing beyond that. With Kepler's discovery of elliptical orbits, the problem of epicycles was resolved. What remained was an explanation of the daily rotation of the celestial sphere around a tiny central Earth, which seemed impossible to explain. However, the correct explanation is that the distant stars do not make a daily revolution around the earth. Instead, only the lights from the distant stars make a complete revolution around the earth diurnally, incidental to its journey towards the more massive sun, according to the universal law of gravitation. This inference supports Newton's corpuscular theory of light. Light, just like any other substance, is material in nature and obeys the law of gravitation. Light beams can not only bend but curve. Therefore, while the distant stars are fixed, lights from these stars make a complete revolution around the central earth in pursuit of the massive sun, which itself revolves around the earth. The sun's revolution around the earth is a mere wobble for such a massive body. All stars should be wobbling because of their hot plasma composition.

Space is three dimensional. Therefore, all measurements of space are cubical and of the third degree. Einstein's famous equation therefore must be corrected as $E=mc^3$. This correction eliminates the need for antimatter, since positive or negative values of *m* and *c* give the correct results. We may therefore very well say goodbye to positrons, neutrinos, bosons, and quarks.

Another amazing aspect of correcting Einstein's equation to the third degree is that the discrepancies between quantum and classical mechanics disappear immediately. The micro and macroworlds obey the same laws of motion and gravity. Principles of quantum physics happen to be the same as those of classical mechanics happening in a microworld where plasma or aether occupies all space.

Prior arguments naturally should lead us to formulate a unified theory that would unite not only quantum theories with classical mechanics but all sciences—material sciences and life sciences. None of these ideas are new. Mankind has contemplated on them and used them for centuries. Mankind has looked at the sky for thousands of years and has found it to be uniformly distributed in all directions. Earth has been found fixed at the center of the sky and all heavenly bodies, including the closer ones, the sun, and the moon, have uniformly revolved around this central Earth.

My version of the grand unified theory is now presented in three ways, borrowing three sets of terms used by Newton (written in italics):

1. During any defined interval of time, in any defined space, for any defined observer, there is only one *absolute* value; all other values are *relative*.

2. During any defined interval of time, in any defined space, for any defined observer, there is only one *true* value; all other values are *apparent*.

3. During any defined interval of time, in any defined space, for any defined observer, there is only one *mathematical* value; all other values are *common*.

Supporters of Relativity Theories missed out on discovering a unified theory because of three main reasons. (1) They rejected aether which is the fourth state of matter (plasma) pervading all space and keeping our universe stable. (2) They also erroneously combined two independent variables—time and space—into a fictitious model called space-time continuum. They called time a fourth dimension without any evidence to support such a theory. (3) Perhaps the Himalayan blunder that relativists made was to believe that there is no difference between a state of rest and a state of uniform motion in a straight line (linear). This erroneous belief had its origin with the Copernican heliocentric solar system which believes that there is no difference between Earth revolving around the Sun or the Sun revolving around the Earth.

Quantum theorists failed to unite quantum with classical mechanics because of their belief in the Copernican heliocentric model of the solar system in which the earth is compared to an orbiting electron instead of a central proton; and the sun is compared to a central proton instead of an orbiting electron. Quantum theorists also failed to define the observer (who is looking?) which makes all the difference.

Both quantum physicists and relativists failed to define the observer. As we have seen in the first discourse, different observers will be seeing the same natural phenomenon differently. Each observer makes a true statement about what he or she is observing. These statements vary based upon the time and place of observation.

Classical mechanics, quantum, and relativity are all combined into a single model when we accept and adhere to the following principles.

1. There are only three dimensions of space—linear, lateral, and vertical.

2. There are six primary directions related to these three dimensions even though we may increase this number at pleasure. These are front and back for linear, left and right for lateral, and up and down for vertical, for every observer.

3. Time is absolute and one-dimensional.

4. Space is absolute and three-dimensional.

5. Matter exists in four states: solid, liquid, gas, and plasma.

6. Space and time are independent variables while matter is the dependent variable.

7. Gravity is the only force in the universe which manifests itself as attraction (centripetal force) and repulsion (centrifugal force).

8. When one observer experiences centripetal force, another observer experiences centrifugal force about the same phenomenon at the same time.

Space and time are exponential by nature but they are two independent variables. Matter is the dependent variable, dependent on both time and space. Since time is an independent variable, space-time continuum is a wrong approach in the search for a unified theory. Time cannot be plotted against space as a fourth dimension. The proposition that time is a dimension is preposterous. Einstein went wrong with this approach and both his theories of relativity are unscientific with major flaws in his arguments.

An explanation of gravity also offers a clue to the puzzle of "Olbers' paradox". We know that there are billions and billions of stars out there and still, at night, we don't see the night sky bright like daytime. Why is this so? The reason is that all the light from those stars are falling on to the sun behind the earth according to the universal law of gravitation because of the huge mass and volume of the sun. Only a tiny fraction of that light falls on Earth because, in comparison to the universe, Earth is only a tiny point. This observation also proves that light is corpuscular matter that obeys the law of gravitation.

Twentieth-century physicists failed to adhere to the strict principles of scientific reasoning in formulating their theories. Theories are to be formulated to explain natural phenomena and not the other way around. Einstein gave up aether theory in order to defend his theories of relativity. Then he made a series of erroneous exceptions to natural laws simply to defend his pet theory even though experimental results tell a different story. Michelson-Morley experiment is a typical example. In this experiment, light was found to travel with the same speed in all directions on the surface of the earth. If the earth were moving, these readings should have been different. A correct interpretation of this experiment is that aether (plasma) exists; but Earth does not move. None of the 20th century physicists dared to make this right inference from the Michelson-Morley experiment. This was the main reason why quantum and relativity theories remained enigmatic and more and more puzzles and paradoxes crept into physics. Time dilation and length contraction are two examples of science fiction writing that any intelligent man or woman would consider voodoo science or science fiction. Even bright students find physics incomprehensible and lose interest in it too quickly.

SUMMARY

My view of a unified theory is stated in three ways using Newton's terminology. Different observers perceive the same phenomenon differently based on their spatial orientation. A meaningful unified theory can only be developed when we include the observer and the object of observation into the model along with strict definitions of time and space. No two observers may agree on the status of a phenomenon, but that does not prove that either one is wrong or right. Each individual should form his or her own theory of unity as part of his or her own personal philosophy.

READER'S NOTES

BOOK II

YOUR DISCOVERY OF THE GRAND UNIFIED THEORY

PREFACE

If you have finished reading Book I; and have written down your questions; you are ready to read Book II for a full comprehension of the Grand Unified Theory. Book II deals with the fundamental principles behind the unified theory, uniting classical mechanics with quantum theories. Chapter 1 discusses the mathematical principles behind the unified theory. Chapter 2 explains the physics behind the unified theory, followed by chemistry, botany, and zoology behind the unified theory. The last chapter is a discussion of the philosophical principles behind the unified theory. Again, it is important to read these chapters in chronological order because there are some new mathematical principles that I have introduced in chapter 1 that are necessary for understanding quantum physics from a classical foundation. I have aptly titled Book II "Your Discovery of the Grand Unified Theory" because when you finish reading this book, you will understand this theory for yourself.

—Arkay Nair
Manchester, Connecticut
August 23, 2013

MATHEMATICS BEHIND THE GRAND UNIFIED THEORY

Mathematics is an essential tool for the study of physical phenomena. However, blind reliance on mathematics, inconsistent with experimental results and scientific reasoning, would lead the scientist into wrong conclusions.

In order to adapt mathematics as a tool for understanding physical phenomena, the following axioms are proposed in addition to the principles of Euclidian Geometry. Mathematicians are welcome to prove or disprove these propositions.

1. The Real Number Line is an arc of the Real Number Circle. The numbers ascend from zero to infinity on the right side as positive numbers and then descend from infinity to zero as negative numbers on the left side. (see Figure 1-1)

2. The concept of infinity is absolute. Absolute infinity stands for + and – infinities. These are not two separate entities but they are the two sides of the same entity. Positive numbers begin with zero and

end at infinity. Negative numbers begin with infinity and end at zero.

3. A circle is a parallel line, parallel to a point called the center of the circle.

4. All parallel lines meet themselves at the origin.

5. For scalar quantities, + sign means addition and - sign means subtraction.

6. For vector quantities, + sign means increasing distance of the object from the observer and - sign means decreasing distance of the object from the observer. If the sign is positive, the object is moving away from the observer. If the sign is negative, the object is approaching the observer. The observer is always assigned zero velocity in order to determine the true motion of the object of observation.

7. For scalar quantities, negative values are virtual. They have no separate existence. They are just mirror images of the positive values.

8. For vector quantities, both positive and negative values are real. They represent the spatial relationship between the observer and the object of observation.

9. Euclidian Geometry is the appropriate tool for representing physical magnitudes.

10. Orbital motions, whether circular or elliptical, are displacements to the origin. Therefore velocities of orbital motions are always zero (see Figure-1-2).

11. In the three-dimensional Cartesian coordinate system, the z-axis is the Observer axis, always occupied by the observer.

12. i = I1I. The square root of negative one is absolute one.

I shall now explain each one of the above axioms as they apply to the measurements of physical phenomena.

THE REAL NUMBER LINE

Traditionally, the real number system has been represented as a straight line beginning with negative infinity and ending in positive infinity with zero as the midpoint. This representation suggests two infinities which is absurd. Infinity is an absolute concept and we cannot attach a number or sign to it. This problem is easily solved by considering the real number line as an arc of a Real Number Circle as shown in Figure 1-1.

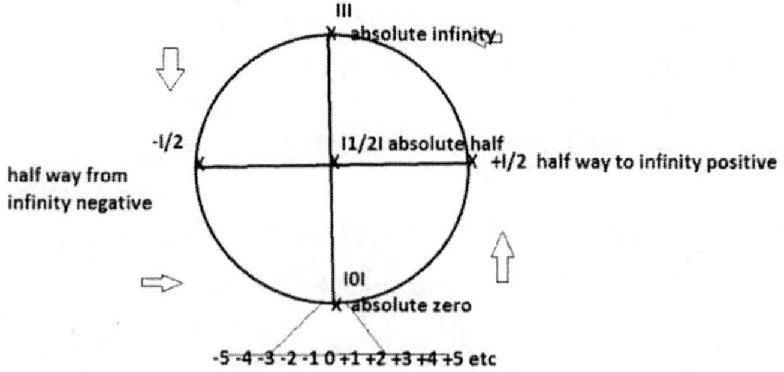

Figure 1-1. Conventional real number line is only an infinitesimally small arc of the real number circle.

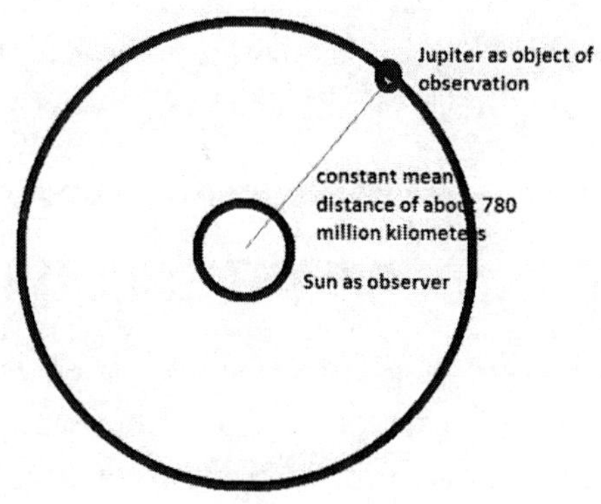

Figure 1-2. orbital motions are displacements to the origin. Velocity is zero.

Figure 1-1 shows the real number line which is actually an infinitesimally small arc of the real number circle (1/ infinity). When dealing with finite numbers; we can only see a very small portion of the real number circle because; the concept of infinity is so immense and absolute. The best way to illustrate this concept is to draw a real number line on the ground, say, at the equator. Stretch the line in both directions and we see that they meet at the opposite side of the earth. The concept of a real number circle becomes highly useful in explaining physical phenomena as we will see in the following chapters.

THE CONCEPT OF INFINITY AS ABSOLUTE

Absolute measures have no attributes. We cannot say one infinity, two infinities, or a thousand infinities. Attributes like negative infinity or positive infinity are also absurd. In Figure 1-1, the y-axis represents the absolute number line, the vertical diameter of the real number circle stretching from absolute zero through absolute half to absolute infinity. The horizontal x-axis represents the mean values or midpoints, halfway from infinity and halfway to infinity. The midpoint of the x-axis is occupied by I1/2I (absolute half). This point is also the midpoint of the y-axis as well as the center of the real number circle.

A CIRCLE IS A PARALLEL LINE, PARALLEL TO A POINT CALLED THE CENTER OF THE CIRCLE

Euclid had been criticized as vague about his parallel lines postulate. However, by using the concept of a number circle, this difficulty is easily resolved. A line drawn parallel to a point forms a circle. The real number circle is a parallel line, parallel to its central point which is the absolute half point in Figure-1-1.

ALL PARALLEL LINES MEET THEMSELVES AT THE ORIGIN

While a line drawn parallel to a point forms a circle, a line drawn parallel to another line also forms a circle. The equator is a line drawn around the earth, equidistant from the poles. All lines of latitude are drawn parallel to the equator and they all meet themselves at the origin.

FOR SCALAR QUANTITIES, + SIGN MEANS ADDITION,–SIGN MEANS SUBTRACTION

Scalar quantities have magnitude only. Positive values are real while negative values are virtual. +5 apples are real while -5 apples do not exist.

FOR VECTOR QUANTITIES, + SIGN MEANS INCREASING DISTANCE OF THE OBJECT FROM THE OBSERVER, – SIGN MEANS DECREASING DISTANCE OF THE OBJECT FROM THE OBSERVER

Vector quantities have both magnitude and direction. While speed is a scalar value, velocity is a vector value. If the velocity is positive, the distance between the object and the observer should be increasing. If the velocity is negative, the distance between the observer and the object of observation is decreasing. For an observer standing on a railway platform, a passing train has negative velocity while approaching the platform, which reaches zero for an instant while passing the platform and turns positive as it leaves the station. Even though the train has been moving at a steady speed of 50 miles per hour without stopping at the station, for the observer on the platform, the velocity of the train is –50 miles per hour until the moment it reaches the station and changes to +50 miles per hour at the instant the train clears the platform.

NATURE OF SCALAR QUANTITIES

Scalar quantities are real, countable objects. Examples are the number of apples in a basket or the temperature of an object in degrees Kelvin. If there are a dozen apples in the basket, you can only take away twelve and nothing more. The same way, you can reduce the temperature of an object down to zero degrees Kelvin but not anymore. You can reduce the speed of a train to zero miles per hour but not anymore.

FOR VECTOR QUANTITIES, BOTH POSITIVE AND NEGATIVE VALUES ARE REAL

The sign of a vector value indicates its direction. Conventional terms like North, West, East, or South are meaningless because they do not specify where the observer is. The only reliable way to distinguish positive and negative values of vector quantities in relation to a stationary observer is by assigning positive number values to an object moving away from the stationary observer and negative number values to an object approaching the observer. For an earthling, a ray of light from the Sun falling onto Earth has -c velocity (-300,000 km/sec.). A ray of light reflected back from the earth has +c velocity (+300,000 km/sec.) from the same observer's point of view.

EUCLIDIAN GEOMETRY

Euclidian geometry is quite sufficient for all measurements of physical phenomena. Euclid's plane geometry applies to all measurements in two dimensions. Euclid's solid geometry applies to all measurements in three dimensions, whether they are spheres and cubes, atoms to crystals, and stars and galaxies which occupy three dimensional space. Non-Euclidian geometry is rejected as unnecessary for physical measurements.

ORBITAL MOTION DEFINED

Scientists generally use the terms orbital speed and orbital velocity interchangeably as if they are the same. This is a major error that has affected the progress of physics for the last four hundred years. Velocities of orbits are always zero even though the speed may vary from about 48 km/sec. for Mercury to about 13 km/sec. for Jupiter. Velocity is different from speed since we have to take into account the direction of the object's motion. According to our new definition of velocity as stated previously, velocities have positive values only if the distance between the object and observer is increasing; and negative if the distance is decreasing. In orbital motions of the planets around the sun, there is no decrease or increase in distances from the sun (the observer). Therefore, orbital motions, whether circular or elliptical, have zero velocity (see figure 1-2).

CARTESIAN COORDINATE SYSTEM

By assigning the z-axis of the Cartesian coordinate system to the observer, we can get many useful results in the study of physics. In Figure 1-2, Sun, the observer, occupies the position x = 0, y = 0, z = 0. Jupiter's orbit around the sun has zero velocity because the mean distance between the sun and Jupiter is always the same; about 780 million kilometers.

i = I1I

The square root of negative one is absolute one. Absolute one is defined as plus and minus one. Therefore, by multiplying absolute one by itself, we get two solutions, +1 and -1. Symbolically, this is represented as I1I (absolute one). This solution may be extended to all complex numbers and quaternions. The sum of these solutions: +1 -1 = 0. Therefore, zero is also a solution for i. Depending upon the specific problem that the scientist is attempting to solve, i may have any one or more of these values: i = +1, i = -1, i = I1I, or i = 0.

With the help of our two new powerful tools, the Real Number Circle and a new definition for velocity, let us now study physics with a fresh perspective in the next chapter.

READER'S NOTES

PHYSICS BEHIND THE GRAND UNIFIED THEORY

EMPEDOCLES OF ACRAGAS

Empedocles was a fifth century BC Greek natural philosopher who is credited to have first postulated that matter exists in four states, which he called earth, water, air, and fire. Substituting the word "solid" for earth, "liquid" for water, "gas" for air, and "plasma" for fire, as they are, we see that Empedocles was the first physicist who realized the true nature of matter existing in four states—solid, liquid, gas, and plasma. Empedocles also spoke about two forces which he called love and strife. Substituting the word "attraction" for love and "repulsion" for strife, we see that these two forces are manifest in the theory of gravitation.

The scientific world paid little attention to this theory because of the definition of the term "element" that Empedocles used in referring to matter. Everyone today knows that there are ninety-two natural chemical elements from hydrogen to uranium and therefore Empedocles must be clearly wrong! However, all scientists would certainly agree that there are only four states of matter—solid, liquid, gas, and plasma. Physicists also dismissed Empedocles as a mere poet when he used the terms "love" and "strife" to describe gravitational forces of attraction (centripetal force) and repulsion (centrifugal force).

PTOLEMY AND HIS MATHEMATICAL COMPILATIONS

In 2nd century AD, Claudius Ptolemy, an Alexandrian astronomer and mathematician, developed a geocentric model of the universe in which Earth is central and stationary and all celestial bodies revolved around the earth. Ptolemy combined his observations with those of earlier astronomers like Hipparchus, who is considered the founder of trigonometry. The Ptolemaic model was practical and useful for the measurement of time in days, months, seasons, and years in agriculture, commerce, and government.

COPERNICAN MODEL OF THE UNIVERSE

Nicolas Copernicus, an eminent mathematician and astronomer of the 16th century, was dissatisfied with Ptolemy's geocentric model of the universe. His reasoning was that it is easier for the earth to rotate on its axis than have the whole universe revolve around the earth. Copernican heliocentric model relegated earth to "just another planet orbiting the sun," third in place among six. This model is an excellent mathematical model. However, Copernicus did not have any empirical proof to support his theory. We must wait until such time when we have conclusive empirical proof before accepting this mathematical model as the true physical model.

TYCHO BRAHE AND HIS GEO-HELIOCENTRIC SYSTEM

Tycho Brahe rejected Copernican heliocentric system because of the heaviness (density) of the earth. He would accept the Copernican model as a mathematical model only, useful for making calculations but devoid of physical reality. Tycho Brahe proposed a geo-heliocentric model in which the moon and the sun revolve around the earth as observed, but all other planets revolve around the sun. This model is an improvement of the Ptolemaic system; agrees with all physical laws and is fully supported by observed results.

NEWTON'S LAWS AND CLASSICAL MECHANICS

Isaac Newton laid the foundations of classical mechanics through his three laws of motion and the universal law of gravitation. Newton developed mathematical formulas for

Galilean mechanics and Kepler's laws of planetary motions. Newton conducted his own experiments to confirm his theories on motion and gravity and published his results in his famous book *Mathematical Principles of Natural Philosophy*. Newton also made significant contributions to optics in which he proposed a corpuscular theory of light, which means that light is made up of corpuscles—extremely minute particles. Newton's Corpuscular theory of light should therefore be considered as the beginning of atomic theories and quantum physics.

NIELS BOHR ATOM

Building upon the quantum hypothesis of Max Plank and emerging ideas on the structure of the atom by Ernest Rutherford and his team of experimental physicists, Niels Bohr developed a model of an atom with a central nucleus and orbiting electrons around this nucleus. J. J. Thomson had earlier discovered the electron in 1897. Rutherford identified the nucleus of a hydrogen atom as a proton; and later, James Chadwick would discover the neutron in 1932. Bohr's model proposed in 1915 worked well for a hydrogen atom because it has only one proton and one electron. This model has been compared to the solar system with the sun corresponding to the nucleus and the electrons corresponding to the planets. The orbiting electrons are supposed to absorb and emit energy continuously. In the Copernican model of the universe, this does not happen because the sun, which is the source of all energy, is compared to the nucleus. However, if we use the Tychonic model in which a dense earth is the nucleus and a light gaseous sun corresponds to the electron, everything falls into proper place. The sun is continuously absorbing and emitting energy and is seen orbiting around a dense but tiny central earth in a well-defined orbit perfectly agreeing with the demands of the quantum theory. We will continue this discussion in the following chapters.

FRITZ ZWICKY AND DARK MATTER

Fritz Zwicky was a scientist of extraordinary vision who predicted the existence of dark matter based on his calculations of the observed and predicted masses of matter in the universe. Perhaps this missing dark matter is plasma (Empedocles' fire) which is invisible at the low temperatures of deep space (2.7K), occupying all so-called empty space. That would certainly account for the 90% or more of missing matter necessary to balance our equations for a stable and balanced universe in equilibrium; that is our observed reality when we look up at the heavens.

EINSTEIN'S ERRORS

Two experimental scientists, Albert Michelson and Edward Morley, performed several experiments to measure the speed of light and its behavior in an aether medium. In all these experiments, they found that the speed of light is the same in all directions on the surface of the earth. If the earth was moving, they should have obtained different values for the speed of light in different directions on the surface of the earth according to all known laws of physics. Michelson Morley experiments are the most conclusive proof that Earth has no motion and the Copernican heliocentric model is not a physical reality. Einstein, as well as all other twentieth century scientists, failed to reach this correct inference. Einstein instead started writing science fiction as special relativity and general relativity by creating exceptions to natural laws as follows:

1. Even though light behaves as a wave, it does not need a medium unlike water waves or sound waves (rejection of the aether/plasma theory).

2. There is no difference between a body at rest and a body in motion (rejection of the notion of inertia).

3. The mass of a body increases as it speeds up (pure science fiction).

4. The length of a body decreases as it speeds up (more science fiction).

5. Time is defined as the position of the hands of my clock (failed to define time properly).

6. My thought experiments are correct or else I feel sorry for the good lord!

7. I would rather trample on Newton instead of standing on his shoulders for a better vision of the universe!

8. Commonsense is nonsense!

Einstein's errors stated above have set us back by another one hundred years in physics while all other branches of science, especially chemistry, have been progressing very well.

MATTER-TIME CONTINUUM AND THE ARKAY CERTAINTY PRINCIPLE

The concept of a space-time continuum as conceived by Einstein is fundamentally wrong. Space and time are two independent variables. Matter is the dependent variable which may be in one place or the other depending on the time you observe it. If you ask me where I was yesterday and I answer, "I was in London yesterday," that is a true description of matter-time-space dependency. I (matter) was in time (yesterday's date) in London (space). If you ask me, "Where was London yesterday," I would laugh at your stupid question! Let us therefore explore how the matter-time continuum principle would help us understand the universe better without having to rely on fictitious concepts of time dilation and length contraction.

Space is uniform everywhere. It is matter that gives space a label and identity. Matter exists in different regions of space at different times in different forms. The function of space is to contain matter and provide matter with an address. This address however changes with time. Einstein went wrong on his first principles when he bundled space with time (two independent variables), instead of matter (dependent variable) with time (independent variable). In the framework of the matter-time continuum, the Heisenberg uncertainty principle can be replaced with my certainty principle: that "for any defined value of time (duration), you can be sure of the exact spatial position of an object of matter."

The human eye can retain a visual image only for 1/24th of a second. A photograph on the other hand retains an image indefinitely. It is quite obvious that here; the limiting factor is the observer and not nature. Let us think of a superman who can retain visual images indefinitely. Such an observer, on observing the sun from Earth over a twenty-four-hour period, will see that the sun is actually a ring around the earth like a doughnut of 1 AU (astronomical unit) radius. Let's call this image a true "Sun-Day." For the same observer, a 'Sun-Year' would look like a spring or slinky of 365 loops (or 365 stacked doughnuts) in the sky stretching from tropic of Cancer to tropic of Capricorn. In general, Matter-Time-Continuum plots matter against time giving us a true picture of the universe. Heisenberg's uncertainty with the simultaneous position and momentum of a particle becomes a certainty when matter, a dependent variable, is plotted against time, an independent variable.

During the life of Sir Isaac Newton, there were only five known planets, not including Earth. Newton's calculations of the center of gravity of the solar system with the available data of his time showed him that the center of gravity of the solar system may very well be within the sun itself. Today, we know that Newton's calculations should be corrected for new data we have gained about three more outer planets—Uranus, Neptune, and Pluto, not mentioning the asteroid belt, comets, Kuiper belt, and even the Oort cloud. The size of the solar system has more than doubled since Newton's times. New calculations will certainly show that Earth is the barycenter (center of gravity) of the solar system. We cannot blame Newton for this error; but all twentieth century scientists clearly missed out on this crucial evidence that establishes Earth at the center of gravity of the solar system. This is the reason why life has developed only here on Earth. It is therefore reasonable to hypothesize that life develops only on stationary planets like Earth, at the center of gravity of any star-planets system. We must therefore rename Earth as a stationary planet at the gravitational center of the solar system. Any search for life anywhere else in the universe should be directed towards finding stationary planets among other solar systems. Shown below are my sketches of the above concepts which are not to scale.

Figure 2-1. Matter-Time Continuum- Sun-Second

observer- an earthling
object- the sun
duration- 1/24th of a second.

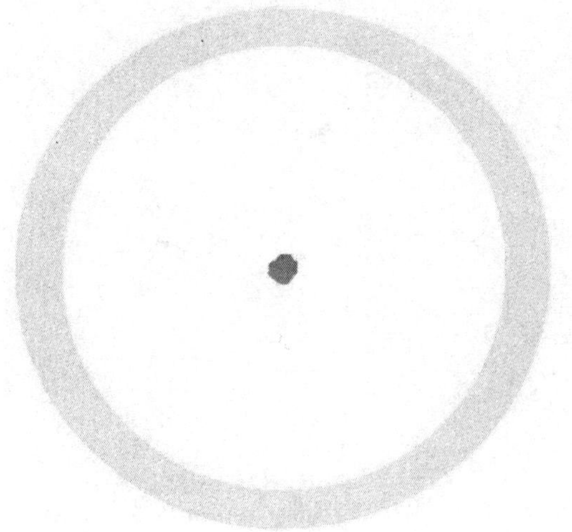

Figure 2-2. Matter-Time Continuum- Sun-Day

observer- Earth
object- the sun
duration- 24 hours

Figure 2-3. Matter-Time Continuum- Sun-Year

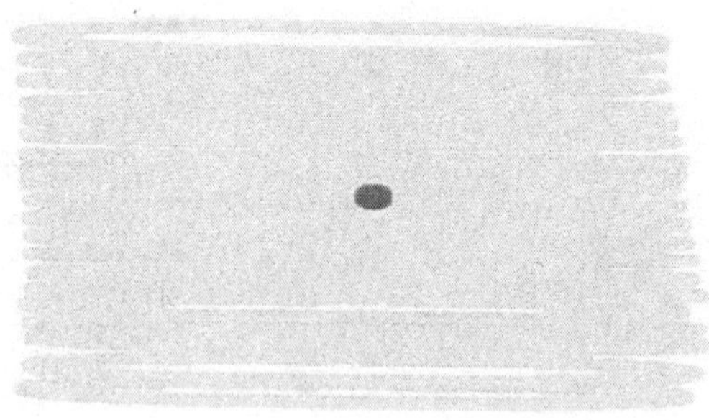

observer- Earth
object of observation- the sun
duration- one year (365.25 days)

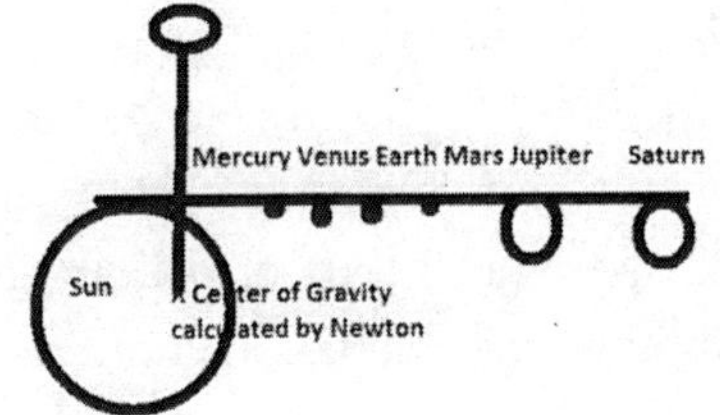

Figure 2-4. Newton's calculations based on a six planet system showed the center of gravity of the solar sytem within the sun itself.

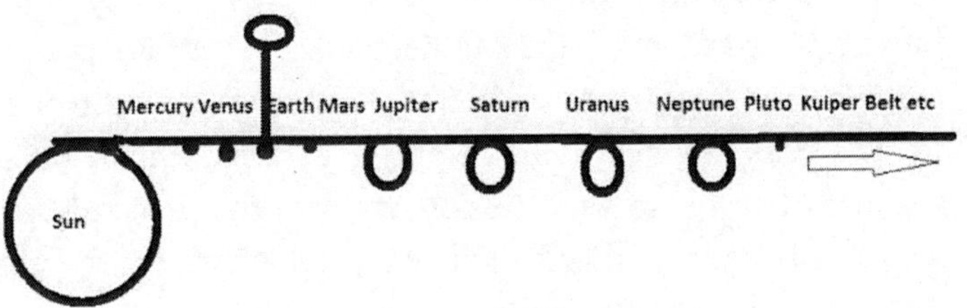

Figure 2-5. A modern day calculation will show that Earth is the barycenter of our solar system as the size of our solar system has more than doubled since Newton's times.

E = MC³

The equation that relates energy with matter, attributed to Einstein, $E=mc^2$ is an incomplete equation. Einstein himself admitted that he needed a cosmological constant in order to balance his equation for a stable universe. This cosmological constant is *c*, the speed of light in deep space plasma medium. The speed of light in deep space is a constant because it is the ultimate terminal velocity matter can achieve. Imagine a hydrogen atom beginning to fall from Proxima Centauri towards earth. The initial speed of zero is steadily accelerated exponentially at 9.8 meters per second until a speed of 300,000 kilometers per second (c) is reached according to the universal law of gravitation. Further acceleration does not take place because the plasma medium resists further acceleration, just as a skydiver in freefall will not accelerate more than 200 km/hr. or so because of air resistance. The constancy of the speed of light in deep space is further conclusive proof that aether exists which is plasma hydrogen, the most abundant element in the universe. Einstein's equation should therefore be corrected as follows:

$E = mc^2 \times c$ (the cosmological constant) ... $E = mc^3$.

In nuclear explosions, half of this energy is the yield. The other half of energy is implosive, resulting in the formation of fission products (lighter elements and huge amounts of neutrons). The yield from any nuclear reaction should therefore be $\frac{mc^3}{2}$. The other half of the energy is directed inwards forming fission products, mainly lighter elements and neutrons. A classic example is the breakup of Uranium 235 into Barium and Krypton with half the energy, $\frac{mc^3}{2}$, available as the yield from the explosion and the other half of energy directed inwards to form fission products, Barium, Krypton, and neutrons. This is an illustration of Newton's third law of motion—For every action there is an equal and opposite reaction. We will continue this discussion in the following chapters.

Space is three-dimensional and multi-directional. What this means is that there are only three dimensions for space even though you may have as many directions as you please. Length, width, and height are the three dimensions of space while north, east, west, south, up, and down are the primary directions. You may increase the number of directions at pleasure such as northeast, southeast, southwest, northwest, etc., but the dimensions cannot be increased more than three. Any idea of dimensions more than three is ridiculous that should be rejected. We may name the three dimensions linear (front and back), lateral (left and right), and vertical (up and down) with respect to a defined observer, each forming right angles with respect to the other two. In a three-dimensional Cartesian coordinate system, the z-axis forms the linear dimension (front and back of the observer), the x-axis forms the lateral dimension (left and right of the observer), and the y-axis forms the vertical dimension (up and down relative to the observer). With this system of coordinates, matter can be plotted against time for any defined observer with full certainty.

CONCLUSION

Physics has lagged behind other branches of science because of certain errors that crept into it during the last four hundred years. These errors started with the acceptance of the Copernican heliocentric model of the universe without solid empirical proof to support his theory. Earth is the barycenter of the solar system based upon our present-day knowledge of the size of our solar system. Stationary earth is therefore different from all other planets which are in perpetual motion. Only stationary earth supports life. Conclusive empirical proof for a stationary earth came with Michelson-Morley experiments. These experiments showed that the speed of light is the same in all directions on the surface of the earth. If the earth were moving, those readings should have been different. It is up to us, the twenty-first century scientists, to correct the mistakes of the past generation and bring physics to the forefront of natural philosophy. On a cheery note, there is plenty of work to do for all physicists and there are many exciting things waiting to be discovered!

READER'S NOTES

CHEMISTRY BEHIND THE GRAND UNIFIED THEORY

Chemistry is the branch of science that deals with the composition, structure, and properties of substances. In this chapter, we will explore the contributions of chemistry to the development of the unified theory as a sister science to physics, mathematics, and life sciences. One important distinction between physics and chemistry is that chemists do not rely on thought experiments unlike theoretical physicists. Chemists strictly follow the methodology of "experiment, observation, and inference."

Carefully conducted experiments and observations have led chemists to reach the following inferences about the structure and function of matter with which our universe is made of.

1. A substance has mass which is its inherent quality.
2. A substance has weight which is the gravitational attraction exerted by Earth on the substance.

3. There are different kinds of substances; but all substances are made up of atoms.

4. Atoms are made up of three fundamental units: protons, electrons, and neutrons.

5. Physical properties of substances are their characteristic qualities.

6. Chemical properties of substances are those properties that convert substances into other substances through chemical reactions.

7. There are four states of matter—solid, liquid, gas, and plasma.

8. Matter cannot be created or destroyed.

We shall now explore these principles of chemistry as they apply to the development of a unified theory.

MASS AND WEIGHT

Mass is a scalar quantity representing the quantity of matter in a substance. Weight of an object is defined as the force of gravity Earth exerts on an object. Weight is therefore a vector quantity of mass directed towards the center of the earth. This is the fundamental difference between mass and weight. Among the elements in the periodic table, the first two elements, hydrogen and helium, have negative weights at normal temperature and pressure. Because of their negative weights, hydrogen and helium are only found in outermost regions of the atmosphere and beyond in deep space. Deep space is not empty. Hydrogen and helium occupy deep space in plasma state at a very low temperature and pressure. This is the dark matter physicists are still searching for.

Weight is a vector quantity with magnitude and direction. All elements except hydrogen and helium have positive weights at normal temperature and pressure. Nitrogen, the next gaseous element in the periodic table after hydrogen and helium, may be assigned zero weight when oxygen, the next gaseous element in the periodic table, may be assigned a slightly positive weight denoted as "0+" weight. Compounds like carbon dioxide and water vapor also have 0+, slightly positive weights, and they occupy the lower regions of the atmosphere.

The weight of an object is dependent on its distance from the center of the earth increasing or decreasing proportionately. An astronaut weighing sixty kilograms on Earth weighs zero kilograms in the International Space Station and negative ten kilograms on the moon. Since weight is a vector quantity, the weight of the astronaut on the moon is a negative value (-10) directed towards the center of the moon, away from the center of the earth.

MATTER AND MOTION

Substances are made of atoms. Atoms are made of protons, electrons, and neutrons. A simple hydrogen atom has a proton as its nucleus with a single electron orbiting the central nucleus. If the orbiting electron collapses into the proton, a neutron is formed. The nucleus of an atom is made up of protons and neutrons. Electrons, equal in number to the protons, orbit the nucleus.

An electron is constantly in motion unless it is fused with a proton forming a neutron. On collision with another neutron, a neutron may break up into a proton and an electron. Protons, neutrons, and electrons have no individuality. Any proton is like any other proton and the same goes with electrons and neutrons. The rest mass ratio of electron to proton is 1:1836. It takes 1836 resting electrons to equal the mass of a single proton. However, a single electron can neutralize the positive electric charge of a proton either by orbiting around the proton or by fusing with the proton to form a neutron. Protons and neutrons exist as spheres while within the nucleus of an atom and electrons behave like strings while in motion around this nucleus. Protons behave like dough or silly putty and electrons look like spaghetti strings.

Neutrons may be considered as the fundamental unit of matter. A neutron splits up into a proton and an electron. An electron may orbit a proton when a hydrogen atom is formed. An electron orbiting a proton and a neutron is the next isotope of hydrogen, deuterium. In tritium, the next higher isotope of hydrogen, a single electron orbits around a proton and two neutrons. An electron in linear motion forms a beta ray. A proton in linear motion forms an alpha ray. A neutron in linear motion forms a gamma ray.

Figure 3-1. A proton may be considered spherical in shape while at rest. An electron may be compared to a string like spaghetti. The low mass of an electron is compensated by its momentum, an orbital speed of c, the speed of light in space. The potential energy of a stationary proton is equal to the kinetic energy of an orbiting electron. This is the reason why electrons and protons have equal charge even though their mass ratio is 1: 1836.

Figure 3-2. An electron in linear motion forms a beta ray. An electron may get kicked out of an orbital motion when hit by a neutron. It then moves linearly at speeds up to c while retaining its negative charge. Such an electron is observed as a beta ray. An electron in linear motion may hit a proton and fuse with it to form a neutron.

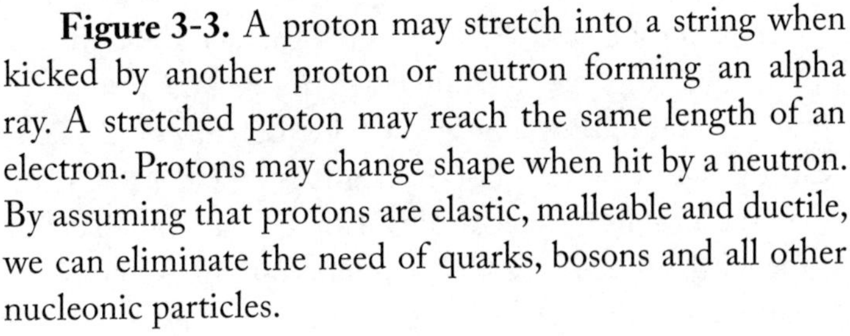

Figure 3-3. A proton may stretch into a string when kicked by another proton or neutron forming an alpha ray. A stretched proton may reach the same length of an electron. Protons may change shape when hit by a neutron. By assuming that protons are elastic, malleable and ductile, we can eliminate the need of quarks, bosons and all other nucleonic particles.

Figure 3-4. A proton and electron (a hydrogen atom) traveling together linearly forms a gamma ray. A hydrogen atom (a proton and an electron) kicked out of the sun may travel together forming a gamma ray. This is the fundamental unit of a light ray which quantum physicists call by the name of 'photon'. Light rays and photons are material in nature, specifically, hydrogen atoms, that have reached the speed of c, the ultimate terminal velocity that matter can achieve, according to Newton's laws of motion and gravitation.

PHYSICAL PROPERTIES OF SUBSTANCES

The first and second elements in the periodic table are gases. These two elements have a natural tendency to escape from Earth because of their negative weights. All the stars including the sun are made up of hydrogen and helium. Hydrogen has three isotopes—protium, deuterium, and tritium. Hydrogen exists on earth mainly as its oxide, water (H_2O). The next four elements are solids under normal temperature and pressure. Nitrogen and oxygen, the seventh and eighth elements, are gases that make up the bulk of our atmosphere. The earth is made up of ninety-two naturally occurring elements from hydrogen to uranium. All these elements are made up of protons and neutrons forming a tiny dense central nucleus and electrons in orbit forming a cloud around the central nucleus. Hydrogen has one proton in its nucleus; uranium has ninety-two.

CHEMICAL PROPERTIES OF SUBSTANCES

Earth is different from other planets in our solar system because life is found to exist only on Earth. The phenomenon of life is a series of complex chemical reactions. Elements react with other elements to form compounds. Oxidation and reduction are the two most commonly found chemical reactions. Hydrogen combines with oxygen to form water. Carbon combines with oxygen to form carbon dioxide. Chemical reactions tend towards equilibrium. At equilibrium, the rate of forward reaction equals the rate of backward reaction. The physiological function of respiration is a typical example of oxidation. Photosynthesis on the other hand is the most important reduction reaction that reverses entropy (disorder) and increases enthalpy (order).

STATES OF MATTER

Generally, substances exit as solids, liquids, gases, and plasma. In solids, atoms and molecules do not move about much relative to each other. In substances of liquid state, molecules are able to move about linearly and laterally in two dimensions. In gases, molecules move about in all three dimensions. Therefore, it is reasonable to hypothesize that solids have one degree of freedom, liquids have two degrees of freedom, and gases have three degrees of freedom. Regarding substances in plasma state, inter and intra-molecular attractions disappear, leaving behind a pool of protons, neutrons, and electrons. These protons, neutrons, and electrons are able to move about freely because of their minute sizes and should therefore be said to have infinite degrees of freedom. In other words, plasma seeps into every nook and corner of space available. There is no space that can escape the penetration of plasma. As these subatomic particles move about in deep space and attain the speed of c (300,000 km/sec.) they exhibit the characteristics of electromagnetic waves. Therefore, it is reasonable to hypothesize that matter and energy are the same; matter in motion and matter at rest. This forms one of the foundations of the Grand Unified Theory.

READER'S NOTES

BIOLOGY (BOTANY AND ZOOLOGY) BEHIND THE GRAND UNIFIED THEORY

Life is an intriguing phenomenon. We know that plants and animals are living things. Some of the most common characteristics of life are metabolism, reproduction, response to stimuli, and power of adaptation to the environment. Metabolism is a series of chemical reactions taking place within the body of the organism that continues until the death of the organism. After death, the body of an organism becomes food for other organisms or returns to nature available for recycling. Reproduction is the process by which an organism produces offspring of its own kind so that life continues to exist through successive generations. Living things exhibit a capacity to adapt to the environment. In our solar system, life has been found to exist only on Earth so far. It is therefore reasonable to hypothesize that life developed on Earth because of its uniqueness compared to other planets. Earth is different from other planets because it is the barycenter of our solar system.

Life exhibits an ability to seek a suitable environment. If the environment is not suitable, life may remain dormant until the right conditions arrive. Life also exhibits an ability to create a suitable environment. Among all the planets in our solar system, life chose to develop only on Earth. It is reasonable to hypothesize that earth is stationary as the barycenter of our solar system offering a solid, immobile ground for the evolution of plant and animal life.

Green plants convert carbon dioxide into carbohydrates in the presence of water which is food for all animals and human beings. Only green plants can reverse entropy created by respiration of animals and human beings. Photosynthesis is the balancing mechanism of life, the only chemical reaction that reverses entropy naturally.

Life is capable of initiating motion. Plants grow their roots into the soil and shoots grow above the ground. Animals are capable of running and birds can fly. Human beings are capable of using tools to enhance motion. Automobiles help man run faster and airplanes help him fly. Rockets help him reach into the space beyond. Life is capable of initiating motion using limbs. Horses run and birds fly using their limbs. Life is also capable of detecting motion using sensory organs. Eyes can detect electromagnetic waves in the visible range. Skin can detect heat. Nose can smell, tongue can taste, and ears can hear. This capacity to initiate motion and detect motion is perhaps the most distinguishing feature of life compared to non-living things like rocks and sand. In other words, life is the observer.

Life is capable of responding to stimuli. A seed on the ground will germinate and grow when the rain falls. A virus will multiply itself into thousands if it can reach into a suitable host cell. Life is always experimenting, observing, and inferring. A hungry man eats food and infers that eating food satisfies his hunger. He repeats this experiment again when he is hungry next time.

Time is the measure of life. While certain bacteria may live only for a few hours, sea turtles are said to live for 150 years or more. Earth is said to be 4.5 billion years old while some scientists have speculated that the universe itself is 13.5 billion years old.

Living things are made up of six essential elements. These are carbon, hydrogen, oxygen, nitrogen, phosphorus, and sulfur. Iron is necessary for animal respiration and magnesium is essential for photosynthesis. Sodium, potassium, and chlorine are necessary as electrolytes and calcium forms the bones of vertebrates. These elements are probably present in all terrestrial planets. However life chose to develop only on earth because it is unique among all terrestrial planets as the gravitational center of the solar system. Comparing Venus with Earth, we see that this twin sister of Earth has all the qualities necessary to develop life. Plants could feast on the carbon dioxide rich atmosphere and reduce it into carbon and oxygen. This would pave the way for later animal life which needs free oxygen for its existence. Earth perhaps started out as Venus is now; or earth may turn out to be like Venus in a few million years. Perhaps our moon made all the difference. Perhaps the Earth-Moon system is the barycenter of our solar system. We do not know. These are my conjectures of possible scenarios.

READER'S NOTES

PHILOSOPHY BEHIND THE GRAND UNIFIED THEORY

Philosophy may be defined as the study of the fundamental nature of reality. Science may be defined as systematized knowledge derived from observation, study, and experimentation. Physics is the branch of science that deals with matter and energy. Matter is defined as that which occupies space and is perceptible to the senses in some way. Energy is the capacity of a physical system to perform work. Potential energy is derived from the position of a body in space relative to other bodies. Kinetic energy is energy possessed by a body by its motion. Human beings use observation and logic as their tools for the study of nature. Nature manifests itself as matter and motion.

Logic is defined as reasoning conducted according to strict principles of validity. Deductive reasoning is reasoning from general statements to reach specific conclusions. If all the general statements are true, the conclusion derived may also be true. Inductive reasoning is general conclusions reached from specific examples. A scientist may perform several experiments and, based on the results obtained, form a theory. We will use both deductive and inductive reasoning in our study of the unified theory that I have proposed in my five discourses in Book I in the same order.

RATIONALE BEHIND A GEOCENTRIC UNIVERSE

In my discourse on a geocentric universe, we found that Copernican heliocentric model is just a mathematical model and not a physical reality. Copernicus reasoned that it is simpler for the earth to rotate on its axis instead of having the whole universe rotate around a tiny central earth. Copernicus himself did not conduct any experiment to prove his theory. It was Galileo who defended the Copernican model based on his telescopic observations and logical reasoning. Galileo's argument goes like this:

1. Jupiter has four moons as satellites and Earth has one moon as satellite.
2. Mercury, Venus, Mars, Jupiter, and Saturn are all planets that revolve around the Sun.
3. Therefore, Earth is also another planet revolving around the Sun.

This argument is logically incorrect. Even though both the major and minor premises are independently true, the conclusion reached from these premises does not necessarily follow. Earth is similar to Jupiter in having a moon. Jupiter is similar to Mercury, Venus, Mars, and Saturn as planets revolving around the Sun. Even though Jupiter is a common factor in both these two premises, they are not sufficient to conclude that Earth is also a planet revolving around the Sun. Let me show you the fallacy of this logic with another example of faulty syllogism:

1. Jupiter is a planet.
2. Saturn is a planet.
3. Therefore Jupiter is Saturn!

It was indeed faulty syllogism that led Galileo into heliocentrism and changed the course of natural philosophy for the next four centuries. Tycho Brahe tried to set things straight (before his early death) by invoking the natural law of inertia, stating that the earth is too sluggish (dense) to move.

William Herschel discovered the planet Uranus in the year 1781, fifty-four years after the death of Sir Isaac Newton. Sixty-five years still later, Neptune was discovered in the year 1846 after its existence was predicted mathematically. Pluto was discovered in the year 1930 and more recently, another dwarf planet, Eris, was discovered in the year 2005. These discoveries have increased the size of our solar system considerably more in mass and volume so that it is quite possible that the gravitational center of the solar system is not within the Sun, as Newton calculated, but on Earth, one astronomical unit away from the Sun. Newton's calculations based on a six-planet system must be revised in the light of these new discoveries. Such calculations will certainly show that Earth is the barycenter of the solar system which is why life developed only here on Earth. It is therefore reasonable to hypothesize that life develops only on a planet that is the gravitational center of a star-planets system.

Accurate measurement of the speed of light was one of the most significant discoveries of the 19th century. Light was found to have a definite speed in any medium. In deep space plasma, light travels at about 300,000 km/sec, designated as *c*. At the other extreme, light travels through diamond (refractive index of 2.4) at a speed of about 125,000 km/sec. It is therefore reasonable to hypothesize that deep space is not empty. Light would reach a speed higher than *c* if deep space was empty according to Newton's universal law of gravitation. Deep space also has a temperature of about 2.7 degrees Kelvin. This phenomenon also indicates that deep space is not empty since empty space will not have any temperature. The ideal gas law, PV = nRT, tells us that when P (pressure) and T (temperature) are very, very low and V (volume) is enormously huge, n (the quantity of matter) has to be a huge number as well, since R (the gas constant) is a constant that does not change. Deep space therefore contains all the missing matter that we need in order to balance our equations for a stable universe.

Albert Michelson and Edward Morley devised an experiment to test the behavior of light in an aether medium. Assuming that Earth revolves around the sun at a speed of about 30 km/sec in orbit, the speed of light should give different values along the direction of motion of the earth and perpendicular (at a right angle) to the direction of motion. The experiment found the speed of light to be the same in both directions. This result was interpreted as a negation of the aether theory rather than a negation of the motion of the earth. Out of the two possible inferences, (1) aether does not exist or (2) Earth does not move, no scientist bothered to analyze the second inference. The Copernican heliocentric model of the universe was so deeply entrenched in the minds of all twentieth-century scientists that no one dared to analyze the second possible inference. It is therefore the task of the twenty-first-century scientists to correct the mistakes of the last generation of scientists and bring physics back to reality. Michelson-Morley experiments gave positive proof that Earth is motionless as observed and that motion belongs to the heavens.

Beginning with Tycho Brahe's improvements of the Ptolemaic model, let us draw a geo-heliocentric model of our solar system based on observed phenomena. In this model we find that:

1. The Sun orbits the earth in about twenty-four hours, similar to an electron orbiting a proton (Sun = electron, Earth = proton).

2. The Moon orbits the Earth in about twenty-four hours and fifty minutes, falling behind the Sun by about fifty minutes each day.

3. At 150,000,000 km radius, the Sun orbits the Earth at a speed of about 10,900 km/sec ($2\pi r/86400$ seconds).

4. This orbital speed helps the Sun remain hot and bright like a meteor.

5. If the Sun was stationary, it would lose heat and brightness and would turn into a black hole.

6. Earth remains stationary at the gravitational center of the solar system.

7. Plant and animal life developed on Earth because it is immobile as the barycenter of the solar system.

8. Only the lights from the distant stars make a complete revolution around the Earth in twenty-four hours, giving us the illusion as if the stars themselves are revolving around the earth.

9. Stars exchange heat and light among themselves, conserving the total amount of energy in the universe.

10. The Sun is constantly receiving heat and light from other stars through the interstellar spaces.

11. The Sun is constantly emitting back heat and light towards its planets and into the interstellar spaces.

12. The stellar exchange of heat and light obeys Newton's laws of motion and the universal law of gravitation.

13. Heat, light, and all other electromagnetic waves are corpuscular matter; more specifically, hydrogen in plasma state.

14. Plasma hydrogen limits the maximum speed matter can attain at *c* (the ultimate terminal velocity) by acting as a resisting medium.

15. If deep space was empty, matter would achieve infinite speed in freefall because of the absence of a resisting medium. Since this does not happen, deep space is not empty. The logical choice of matter that fills all deep space is hydrogen in plasma form at an extremely low pressure and temperature.

There are only four types of observers. They are: Earthling, Earth, Alien, and Sky. Earthling has limited vision from a single point on the surface of the earth. At any time, an earthling cannot see more than in one direction, at the most 120 degrees in angle with the best of peripheral vision. Earth, on the other hand, has a 360–degree vision in all directions towards the sky. As far as the earth is concerned, all heavenly bodies are at a 90-degree angle straight overhead. An earthling observes dawn and dusk while it is always high-noon for the earth! Earth's vision may be compared to the compound eye of a common housefly. An earthling is only a single unit of this compound eye. An alien is an observer at any single point in the sky looking towards Earth. The combined vision of all aliens from all points in the sky is the vision of the sky. It is therefore easy to see that earthlings and aliens have limited vision while Earth has complete vision outwards and Sky has infinite vision inwards.

READER'S NOTES

DISQUISITIONS ON THE STRUCTURE AND FUNCTION OF THE UNIVERSE

The Tychonic model of our solar system is also consistent with atomic and quantum theories. These theories describe an atom as a dense, tiny central nucleus of protons and neutrons with electrons orbiting the central nucleus at very high speeds. Let us look at some possible models of our universe taking examples from both micro and macroworld phenomena.

1. The Deuterium model: Earth = proton, Moon = neutron, Sun = electron.
2. The Water molecule (H_2O) model: Earth and Moon = H_2, Sun = O.
3. The Electromagnetic model of the earth-sun system.
4. A Gravitational model of the universe.
5. A Quantum model of the universe
6. A Biological model of the universe.

7. A Classical Mechanics model of the universe.

Earthlings see the moon and the sun as large bodies when viewed from the earth unlike the planets and the stars which appear as points of light only. In the deuterium model, the earth and the moon together form the nucleus. The sun orbits this nucleus similar to an electron. The sun is constantly absorbing and emitting energy just like an electron does when orbiting its nucleus at a very high speed. Please note that I used the word "speed" and not velocity. Remember that we have already established that orbital motions have zero velocity. The sun orbits the earth-moon system nucleus at a very high speed but with zero velocity. The sun's motion is not linear but orbital. Therefore, its velocity is zero while its orbital speed is about 10,900 km/sec.

In the H_2O model of the solar system, the earth and the moon represent the two hydrogen atoms in a water molecule. The sun represents the oxygen atom with surprising similarity, having eight planets orbiting the sun representing the eight electrons orbiting an oxygen nucleus. It is easy to see that the Tychonic geo-heliocentric model of the solar system is more consistent with quantum and atomic theories.

The Earth and the Sun together resembles a dynamo, electric motor, or an electromagnet. The earth has a metallic core of iron, cobalt, and nickel and the sun is rich in protons and electrons. The revolution of the sun at a high speed around the earth is similar to an electric current that induces magnetism into the earth, making it a permanent electromagnet. The sun's daily orbit around the earth is similar to one loop of the conducting coil. In one year, the sun makes 365 loops around the earth, keeping the earth a permanent electromagnet. If the earth were orbiting the sun, it would not get magnetized to the same extent as it has now. This is also positive proof that it is the sun that is orbiting the earth and not vice versa (See Figure 2-2 & 2-3).

A gravitational model of the universe assumes that there is only one force in the universe which is gravity. Bodies are attracted to each other directly in proportion to their masses and inversely in proportion to the square of the distance between them. The least massive object in the whole universe is an electron followed by a proton. A proton is about 1836 times more massive than an electron. This is the reason why electrons orbit protons forming the simplest of the atoms, the hydrogen atom. An electron in linear motion manifests as the phenomenon of electricity. Since everything else in the universe is much more massive than the electron, the electron is attracted to everything else in the universe—whether it is a proton, a neutron, an atom, or a molecule. Protons and neutrons normally remain stationary within the nucleus of atoms. Protons may be similar to billiard balls and a neutron may be compared to the cue ball. Neutrons are slightly more massive than protons because it includes the rest weight mass of an electron. Therefore, neutrons can kick protons out of a nucleus. An electron that enters the nucleus may fuse with a proton to form a neutron. When a beam of neutrons enter a large atomic nucleus like the uranium nucleus, the neutron beam may split the nucleus into two or more smaller nuclei, just as a cue stick hits the cue ball, which in turn scatter the racked billiard balls in different directions. A proton and an electron fuse to form a neutron; a neutron splits into a proton and an electron when hit by a stream of neutrons with sufficient force. This is the principle of all nuclear reactions.

Quantum theory remained enigmatic because of the defects in its fundamental principles. Whenever a new theory is proposed, it is necessary to reevaluate all old theories in the light of the new theory. Heliocentric solar system, wave theory of light, and rejection of the aether theory did not go well with quantum theories. If we examine quantum theories in the light of a geocentric solar system, corpuscular nature of light and all electromagnetic waves, and the existence of aether as plasma hydrogen filling all available space, we see that quantum theory is in perfect agreement with classical mechanics revised and corrected to include new data regarding the size of our solar system and the speed of light with a limit of *c*. In other words, classical mechanics and quantum theories describe the same universe from two different perspectives, macro and micro models of the universe.

In a biological model of the universe, an electron may be compared to a sperm and a proton similar to an egg. Only one sperm may fuse with an egg when fertilization takes place. The fertilized egg is similar to a neutron impenetrable to any more sperms. A hydrogen atom (protium) may be compared to the marriage of an electron (man) and a proton (woman). A deuterium atom corresponds to the woman getting pregnant, and tritium when she is pregnant with twins!

Having considered all the above models separately, let us go back to our good old classical mechanics model incorporating newer discoveries and theories into this model. Greek philosophers had contemplated on both heliocentric and geocentric models of our universe well before the beginning of the Christian era. However, the geocentric model prevailed because it was consistent with observation and all known physical laws. Ptolemy consolidated his own measurements with those of his predecessors and created the *Almagest* in the 2nd century AD. This model survived for 1500 years until Copernicus revived the old discarded heliocentric system in 16th century AD. Being an eminent mathematician and astronomer, Copernicus created an excellent mathematical model, but this model is not a physical reality. A moving earth would create a drag which should be detectable. Luckily for Copernicus, two other mathematicians from the next generation, Galileo and Kepler, accepted his model and promoted it vigorously. Tycho Brahe was the lone dissenter who was neglected because of his early death. Tycho Brahe proposed an improvement on the Ptolemaic system by suggesting that the sun and the moon orbited the earth but all other planets orbited the sun. This is the true model that we observe every day. Isaac Newton was a genius in both mathematics and physics. He was humble enough to acknowledge the contributions of earlier scientists and benefit from their discoveries in advancing scientific progress. Using the data available at his time, Newton found that the center of gravity of the solar system must be within the sun itself because of the enormous mass and size of the sun compared to the masses and distances of the six known planets of his time. If Newton were alive today, he would be the first person who would endorse my hypothesis that Earth is the barycenter of the present-day solar system.

The immobility of Earth was positively confirmed by the Michelson-Morley experiments which showed that the speed of light is the same in all directions on the surface of the earth. Speed of light is finite, limited to c, about 300,000 km/second. This discovery proves that deep space is not empty. There is something out there that resists the propagation of light beyond c even though it is invisible to human beings. This substance is plasma which is a pool of protons, electrons, and neutrons with a detectable temperature filling all of deep space. Why is this plasma or aether invisible? Aether is invisible because it is the smallest of matter in form below the limit of human observation. Just as we cannot see air, we cannot see aether either! However, deep space has a temperature of about 2.7 degrees Kelvin. Using the equation PV = nRT, we should be able to calculate the pressure of plasma matter in deep space. This would lead to the discovery of the quantity of matter in deep space and the volume of our known universe.

The distant stars are fixed. Only the lights from these stars make a complete revolution around the earth diurnally. Lights from the distant stars are attracted to the sun because of its large mass and size. As the sun revolves around the earth, lights from all other stars are chasing the sun in orbit around the earth. Observing from Earth, this phenomenon looks as if the stars are circling the earth. This is only an illusion. Only the beams of light from the stars circle the earth in pursuit of the sun. This phenomenon proves that light is corpuscular as Newton proclaimed. Light travels between stars obeying Newton's laws of motion and gravitation.

READER'S NOTES

BIG BANG VERSUS BING BANG

Many physicists believe that the universe started with a Big Bang some 13.5 to 14 billion years ago. The theory originated from observing the red shift of light from some distant galaxies which is presumed to confirm a theory of an ever expanding universe. This theory has many flaws in it. I shall list below some of the most prominent errors that I have noticed in the Big Bang theory.

1. All red-shifted galaxies are hundreds of thousands of light years away. Since the velocity of light is finite, at best, we can only say that they *were* moving away from the earth 100,000 years ago. We do not know if they have stopped expanding. We will have to wait another 100,000 years to find out the truth about it.

2. Not all distant galaxies are red-shifted. In fact, there are many blue-shifted galaxies nearer to the earth, indicating that expansion is not uniform like the blowing up of a spotted balloon. This observation should be interpreted as a universe tending towards

equilibrium rather than expanding or contracting. While some galaxies are moving away from the center (wherever that center is) some galaxies are approaching the center. Therefore, at best, the universe is in a steady state of equilibrium just as we have observed over the past several thousand years without any major changes.

3. A Big Bang does not happen without a Big Bing, according to Newton's third law of motion. This means that expansion has to be associated with contraction where matter is condensed to such a high degree that an enormous back hole or neutron star is created as the center. This back hole or neutron star will be the center of the universe from where matter began expanding. The gravitational attraction of such a black hole or neutron star will pull the universe together again perhaps 13. 5 billion years into the future. If contraction starts today, we will not know about it for the next 13.5 billion years. If you keep on inflating a balloon, it must burst eventually.

4. Big bang theorists have nothing to say about what happened before the big bang. It is quite possible that the big bang is cyclical. Today's expanding universe may someday start contracting and finally

reach singularity. This would mark the beginning of the next big bang billions of years into the future.

5. For the above reasons, the Big Bang theory is not superior to any other theory including religious beliefs. It is better to hypothesize on the birth and formation of our Earth, which is finite, rather than that of the universe which is infinite from our viewpoint. I believe that the earth was perhaps formed by a supernova type of explosion of a companion star of our Sun, perhaps 4.5 billion years ago. Binary star systems and supernova type of explosions are common occurrences in the universe. When a star explodes, one half of its mass is condensed into the center and the other half escapes into the space as plasma hydrogen. We may call this phenomenon the "Bing-Bang" in order to stress the importance of the implosion which is equal in force to the explosion as Newton predicted. The central mass eventually cools and condenses into various elements. In the case of our Earth, ninety-two natural elements were successively formed from hydrogen to uranium during this cooling period. Later on, as the cooling continued, these elements reacted among themselves chemically to form compounds.

READER'S NOTES

THE FORMATION OF ELEMENTS AND COMPOUNDS

Binary star systems are quite common in our universe. Even our nearest star, Alpha Centauri, is a two- or three-star system. Supernova type of explosions of stars is also commonly observed in our universe. It is therefore quite possible that our earth was formed through such an explosion of a companion star of our Sun, perhaps 4.5 billion years ago. Since we know that our earth is finite, we may speculate about its birth in a preexisting universe. An explosion is always accompanied by an implosion which is the reason for calling this supernova type of explosion the "Bing-Bang." The Bing-Bang is followed by a cooling period. Hydrogen in plasma state is just a huge pool of protons and electrons in equal number. One half of the matter of this companion star is ejected outwards while the other half is compressed inwards in an equal and opposite reaction according to Newton's third law of motion. A good quantity of the escaping plasma would have accreted into our Sun which became more massive. The compressed matter formed the earth and other terrestrial planets and bodies. What is left over perhaps accreted as the gaseous planets—Jupiter, Saturn, Uranus, and Neptune.

As the earth cooled, protons and electrons started pairing. When electrons collapsed and fused with protons, they became neutrons. As the core pressure of the imploding star increased, more and more elements were formed until ninety-two natural elements were formed that became our Earth. It is interesting to note that lighter elements are more abundant in the earth's crust while heavier elements are more abundant in the core of the earth.

As the cooling progressed, elements started reacting among themselves. Hydrogen combined with oxygen to form water and carbon combined with oxygen to form carbon dioxide. At this stage, Earth would have looked very much like what Venus looks like today. However, Earth had a slight advantage over its twin sister, Venus, as the barycenter of the solar system. While life developed and flourished on Earth, Venus remained barren. The presence of the Moon also perhaps contributed to the stability of the Earth, allowing life to develop here rather than on Venus. It is therefore reasonable to hypothesize that life develops only at the gravitational center of a star-planets-moons system. If a planet or planet-moon system happens to be at the barycenter of a star system, life may perhaps develop on that planet. Perhaps Jupiter may become a star one day in the distant future when our Sun undergoes a supernova type of explosion.

READER'S NOTES

FORMULATION OF A UNIFIED THEORY

Twentieth-century scientists would have succeeded in formulating a unified theory if they had strictly followed the principles of scientific reasoning—Experiment, Observation, and Inference. A theory may be defined as a supposition or a system of ideas intended to explain something based on general principles. A hypothesis is defined as a supposition or explanation made on the basis of limited evidence as a starting point for further investigation. A theory may rise to the status of a law when extensive experimental evidence supports the theory. Newton's laws of motion and gravitation have reached the status of laws as everyday experience proves these laws true without exceptions. At the beginning of the 20th century, Quantum Theory was born with Max Planck's investigations of Blackbody Radiation. Quantum theory deals with the study of matter and energy at subatomic and atomic levels. Human eyes cannot see atoms even with the help of the most powerful microscopes available. Atoms become visible when a certain number of them are heaped together into a pile. As an example, if we mark a dot on a white sheet of paper with a sharp pencil, we are seeing a bunch of carbon atoms (graphite) heaped up. We could make the mark even with a single carbon atom but no man or woman would be able to see it. Similarly, a hydrogen atom in freefall from a star towards Earth becomes a light beam when it reaches the speed of *c* (300,000 km/sec.). This beam may not be visible to the human eye, but when a thousand such beams are bunched together, it may become visible to humans. This is the basic principle of quantum theory. It takes a certain quantity of matter to make it detectable. This limit may vary from matter to matter and from observer to observer.

Niels Bohr proposed a model of the atom similar to the planetary model of the solar system. This model did not get accepted because the Sun was compared to the nucleus and the planets were compared to the electrons orbiting the central nucleus under the Copernican heliocentric model of the solar system. Today, in hindsight, we can easily see that the Tychonic geo-heliocentric model of the solar system perfectly fits with Niels Bohr model of the atom. Electrons are supposed to be moving fast and absorbing and emitting energy in that process. The Sun is constantly moving through plasma, absorbing and emitting plasma while in orbit around the central nucleus. This central nucleus is the Earth (similar to a proton) or the Earth-Moon system (similar to a proton and a neutron). From this observation, we may infer that the Niels Bohr model of the atom is to be accepted and Copernican heliocentric model of the solar system is to be rejected.

Theories of Relativity also ran into trouble owing to the prevailing heliocentric theory of the solar system. Einstein had to make several exceptions to natural laws in order to conform his theories of relativity with observed everyday natural phenomena. Both Galilean and Einsteinian theories of relativity are based upon the belief that the earth is moving but we don't feel it. Galileo and Einstein both forgot that the same first principle equally supports both theories: geocentric and heliocentric solar systems. This argument goes as follows:

1. Relativity theory says that we will not know if the earth is moving or the sun is moving. Therefore, both assumptions are equally possible.

2. If the earth is moving, we should feel a constant wind. Since we don't feel a constant wind, the earth is not moving.

3. Motion is associated more with gases and less with solids. The Sun is a huge ball of gases while the Earth is dense and solid. Therefore, the Sun is moving and not the Earth.

4. Life is found only on Earth in the solar system. Earth is therefore unique in the solar system as the barycenter of the solar system, stable and immobile.

5. Mercury, Venus, Mars, and even our Moon have many impact craters, more than Earth receives, giving us a clue that these planets and our Moon are in constant orbital motion while our Earth is stationary. If the Earth were orbiting, it would also be subject to many impacts that would make existence of life on Earth impossible.

6. Since the times of Newton, no one has bothered to recalculate the center of gravity of our solar system, even though the size of our solar system has more than doubled since Newton's times through new discoveries.

Einstein made a huge mistake when he combined time with space as a fourth dimension. Space and time are two independent variables. It is matter that is the dependent variable, variable with time. There are only three dimensions which are linear, lateral and vertical in relation to an observer. These three dimensions produce six primary directions such as front and back for linear dimension, left and right for lateral dimension, and up and down for vertical dimension. We may increase the number of directions as we please but we cannot find any dimension more than three. Time is not, in any way, a fourth dimension.

We are now ready to formulate a unified theory based on the following observed and verified natural phenomena.

1. Matter exists in four states: solid, liquid, gas, and plasma.

2. Space is three dimensional which are linear, lateral, and vertical.

3. These three dimensions are made up of six primary directions: front and back for linear dimension, left and right for lateral dimension, and up and down for vertical dimension in relation to an observer.

4. Space and time are two independent variables.

5. Matter is the dependent variable, dependent on space and time.

6. There are only four types of observers: Earthling, Earth, Alien, and Sky.

7. Each observer has a distinct point of view different from the other observers.

8. Different observers will be forming different inferences based on each observer's position in space, time of observation and object of observation.

Based on these principles, I have formulated a unified theory which you can read in Book I Discourse 5. The theory is stated in three ways using Newton's terminology as follows:

- During any defined interval of time, in any defined space, for any defined observer, there is only one *absolute* value; all other values are *relative.*
- During any defined interval of time, in any defined space, for any defined observer, there is only one *true* value; all other values are *apparent.*
- During any defined interval of time, in any defined space, for any defined observer, there is only one *mathematical* value; all other values are *common.*

SOLUTIONS TO PUZZLES AND PARADOXES

I shall now attempt to answer several puzzles and paradoxes in physics that are easily solved when reviewed under my unified theory.

1. *Thomas Young double-slit experiment.* In this experiment, light was found to behave as a wave rather than as a stream of particles. When light was passed through one slit only, light behaved as a stream of particles. When light was passed through two slits, an interference pattern was noticed similar to a wave pattern. The reason why light behaves as

particles and waves is that light is the fourth state of matter, plasma. Plasma fills all available space. Plasma is the medium in which light propagates. Plasma is present everywhere, even in deep space. We cannot create a complete vacuum.

2. *"Failure" of Michelson-Morley Experiment.* The experiment was not a failure; only the inferences failed to fit existing theory that earth moves. The correct inference is that Earth does not move.

3. *Cosmic background radiation.* Earth is receiving a constant stream of microwave radiation. Electromagnetic waves are corpuscular. These waves are hydrogen in plasma state or neutrons in linear motion hitting the earth as a result of freefall from the sun and other stars according to the law of universal gravitation (the sky is falling!).

4. *Sun's corona is hotter than surface.* All the light and heat from all other stars are falling on to the sun which is the most massive and voluminous body in our solar system. Sun's surface acts like a perfect convex mirror and reflects back this heat and light back to the stars from where it came from. It is therefore reasonable to hypothesize that Sun is not producing any heat and light of its own. All the light and heat in the universe is preserved through a constant interchange of plasma between stars. All

the stars, including our Sun, are simply reflecting back and forth light and heat according to the universal law of gravitation. This brings us to the next paradox, Olbers' Paradox.

5. *Olbers' Paradox.* Olbers' paradox asks the question "Why is the night sky dark?" The night sky is dark because the earth is only a tiny point in size compared to the rest of the universe. The amount of light that can converge onto a point is very limited. At night, the sun is directly behind the earth. All the light from all the stars are falling on to the massive and voluminous Sun behind the earth rather than on to a tiny earth because of Sun's greater gravitational attraction. Therefore it is reasonable to hypothesize that there is no nuclear reaction taking place in the sun, but merely a reflection back and forth of light and heat between all the stars in the universe.

6. *Why is the sun stable?* If the source of energy from the sun is nuclear reaction, then there is no reason why this reaction cannot progress into an explosion, with the abundance of the nuclear fuel, hydrogen. A better explanation is that there is no nuclear reaction in the sun but merely an exchange of heat and light from all the stars in the universe. Being spherical, the sun acts like a perfect convex mirror reflecting back all the light and heat that falls upon it from

all the other stars in the universe. The earth receives its share of heat and light according to the inverse square law. All other planets and satellites receive their shares as well, according to the same inverse square law.

7. *Why life developed only on Earth?* All the terrestrial planets have elements necessary to form life. However, life developed only on Earth in our solar system. There must be something special about Earth that is unique to Earth only. This has to be its central position as the barycenter of the solar system. Newton's calculations based on a six-planet solar system showed him that the center of gravity of the solar system may very well be within the sun itself. Uranus, Neptune, and Pluto were discovered much later and they are much farther from the Sun than Jupiter and Saturn. If Newton were alive today, he would be the first person to endorse my theory that *Earth is the gravitational center of the solar system and that is why life developed only here on Earth.* We do not know if Earth is unique in the universe, but it certainly is unique in our solar system.

8. *Why is dark matter invisible?* Dark matter is invisible to human beings even with the most powerful of the telescopes. The reason is based on the concept of limit. There is a limit to the extent of human vision both of

the macroworld and the microworld. Dark matter is hydrogen in plasma state, thinly dispersed in space, pervading all available space without exception. Being the smallest of material particles and most abundant, nothing can prevent its penetration and occupation into all available space in the universe. Thus, dark matter is everywhere, but we cannot see it. We feel its presence through cosmic background radiation. We know its temperature in deep space is about 2.7 degrees Kelvin.

9. *The myth of antimatter*. It was Paul Dirac in his attempt to reconcile Quantum and Relativity who proposed a theory that all particles may have an antiparticle with nearly identical properties, except for an opposite electric charge. We have already seen that Einstein's famous equation $E = mc^2$ is incomplete. Therefore, Dirac's defense of this equation is also based on erroneous first principles. When we correct Einstein's equation to the third degree, $E = mc^3$, we see that Dirac's arguments for antimatter are unnecessary. We have already established that c and $-c$ are different only in direction. An electron traveling away from the observer has a velocity of $+c$; while an electron approaching the observer has a velocity of $-c$. In a nuclear explosion, there is an equal and opposite implosion which forms lighter elements and neutrons. When a nuclear mass of

m explodes, $\frac{mc^3}{2}$ is the explosive part of the reaction which is called the "yield" from the reaction. The other half of the energy is directed inwards forming lighter elements and neutrons. This is the correct perspective about nuclear reactions. There cannot be an explosion without a corresponding implosion. It is therefore easy to see that Paul Dirac's antimatter is a myth. Matter cannot be created or destroyed (especially by your evil twin!). Commonsense prevails in the end.

10. *The myth of time dilation.* Time dilation is pure science fiction. Time is the measure of life. When we say Isaac Newton died at the age of 85, the earth is 4.5 billon years old, the half-life of iodine 131 isotope is 8 days, etc., we are talking about the lifespan of matter, whether living or non-living. Time is defined as a measurable period during which an action, process, or condition exists or continues. Clocks are only devices for measuring time, they can never be time itself. Einstein's argument that a moving clock slows down is only the result of the finite speed of light. It has nothing to do with time itself. If you can build a large clock with a bright luminous dial and take it to a distance of one light-hour from Earth after synchronizing it with a clock on Earth, you will see that the distant clock is one hour behind the earth-clock at this new location. The reason is

that you are seeing the hands of the clock as it was one hour ago since light that started from the face of the clock takes one hour to reach an earth observer. If you used a light beam to transport this clock to and from this distance, you will see that the clock's hands are frozen showing the starting time on its journey to the distant station traveling at the speed of *c* on a light-beam ship. The clock's hands turn double as fast on return so that when this clock gets back to Earth; both clocks are again automatically synchronized. When traveling at the speed of light, *c*, time seems to freeze outbound, because the clock is receding from the observer at speed *c* (velocity = +c). Light from the face of the clock is approaching the observer at speed *c* (velocity = –c). Since +c –c = 0, the observer sees the clock face frozen. On return, the clock face will gain back the lost hour. The hands of the clock on return will seem to be running double fast for the Earth-based observer. On reaching Earth, both clocks will show the same time. Time progresses arithmetically. Apparent gain or loss in one way travel is always lost or gained on return travel.

11. *The myth of length contraction.* George FitzGerald and Hendrik Lorentz proposed this theory in order to explain the "failure" of Michelson-Morley Experiment. As we have seen already, Michelson-

Morley Experiment was not a failed experiment. Only the inferences formed from the experiment failed to support theories of relativity. Objects do not contract on speeding up. Instead objects break up at high speeds because of increasing resistance from media, whether it is from air (gases) or from plasma (in deep space). Remember how Comet Shoemaker-Levy broke up into thousands of pieces before impact with Jupiter in the year 1994.

PRACTICAL APPLICATIONS

Perhaps the most important contribution a unified theory offers to the modern-day world is the awareness that energy is a commodity just like corn and cattle. Starlight is stardust which falls more on certain places on earth and less in certain other places. People who live in places where sunshine is plenty should use sunshine for all their heating needs. Solar cookers should do all the cooking. Concentrated solar power should be the source of power for factories and transportation. Lands suitable for agriculture should be used fully to grow green plants. Only green plants can improve our environment by reversing entropy. People who live in temperate climates should make full use of cold weather for all their needs in refrigeration. Solar, wind, and hydropower should be used to reduce dependence on oil and coal. Homes should be built energy efficient with solar, wind, tidal, and geothermal resources fully utilized for local residential needs.

A unified theory based upon geocentric solar system and corpuscular nature of all electromagnetic waves has many practical applications. Since the industrial revolution, man has been consuming more of Earth's resources than he can replace. Human beings are burning more carbon than green plants can regenerate. Judicious use of carbon along with re-plantation of trees would be the first step we must take to improve our environment. We must remember that photosynthesis is the only process that converts carbon dioxide back into carbohydrates. We take it for granted that we can live if we breathe the air around us on this earth. If we breathe the air of Venus or Mars, we won't live long! There are no green plants on those planets to reverse entropy and make human and animal life possible. Mercury, Venus, and Mars are at their maximum entropy level.

Living things are different from nonliving things because of intelligence. We may even hypothesize that life is plasma, the fourth state of matter. Since plasma has access to every nook and corner of the universe, life, which is perhaps plasma, is constantly searching for suitable places to settle and grow. Just as barnacles attach themselves and grow on the hull of a ship and not on the propellers, life is searching for stationary planets like Earth to settle and grow. Therefore, we must protect and preserve our planet since this is the only good real estate we have in our neighborhood! We must use our intelligence to create balance and harmony with nature in order to sustain life here on Earth for a long time. We must remember that even our nearest star is over four light-years away.

In order to protect and preserve our natural resources for many future generations, we must exercise discipline and restraint in our use of these resources. We must be self-sufficient in our use of energy on a current basis. This means that whatever energy we receive from the sun daily should be used for our daily energy needs. Tapping into our reserves should be kept to a minimum and must be replaced.

Carbon sequestration is not necessary. Plants are capable of handling more carbon dioxide since it is not the limiting factor in photosynthesis. In other words, if more carbon dioxide is released into the atmosphere, plants will use more of it to produce carbohydrates. Growing more trees is a better solution than carbon sequestration. We don't want to starve our trees! Further research is needed in this area.

All energies must be expended productively. Exercise machines should be producing electricity rather than consuming electricity. Children should be encouraged to charge their laptops and cell phones while exercising on machines that produce electricity rather than consume electricity. Men and women should boast about the kilowatts of power they produced when exercising rather than about how many calories they burned. In thinking along these lines, the possibilities are endless. Human muscle power should always be used to produce energy so that we can feel good about ourselves as producers rather than as consumers. Strictly speaking, farmers are the only true producers of energy on earth. Everyone else is either consuming, or at best, exchanging energy.

A unified theory also helps us to solve problems in global warming and climate changes. We are using energy as if it is a wave that passes by, leaving behind nothing. This is not true. James Joule's paddlewheel experiment proved that mechanical energy is converted into heat energy as the paddlewheel heated the water on spinning in it. We have thousands of ships on voyage on our oceans spinning their propellers for mechanical motion. Those energies just don't fade away. They are just transferred into the ocean water which in turn contributes a little towards global warming. We have thousands of jet airliners flying around whose jet streams transfer heat to the atmospheric air. Energy is never lost but is always conserved. But for green plants, entropy would proceed to a maximum and life would become impossible on Earth. Since the only way of reversing entropy is through photosynthesis, we must grow more green plants and trees in order to protect our environment.

RESEARCH OPPORTUNITIES

Researchers, teachers, and students have plenty of opportunities for basic research in physics, chemistry, and life sciences once they are able to grasp the essence of this unified theory. Some of the most important ideas are listed below.

1. Calculation of the gravitational center of the present day solar system.
2. Calculation of the pressure of deep space plasma given the temperature of 2.7 K.
3. Calculation of the mass of deep space plasma (dark matter) given temperature and pressure.
4. Calculation of the yield from a nuclear explosion given that total energy $E = mc^3$ and yield $= \frac{mc^3}{2}$.
5. Calculation of the minimum level of CO_2 green plants need to survive.
6. Calculation of maximum level of CO_2 plants can tolerate.
7. All other ideas that my readers can come up with under this unified theory.

CONCLUSION

On the cover of this book, you can see a picture of a sphere within a cube. The diameter of the sphere is the same as a side of the cube with unit length of 1. The sphere is supposed to represent our universe, the universe we can see, feel, and explore. The cube is supposed to represent the infinite universe. The earth and all earthlings are at the center of the sphere which is also the center of the cube. The volume of a sphere is $4/3\pi r^3$. With a radius (r) of 1/2 unit, the volume of the sphere is 0.523598776 unit on my TI calculator. The volume of the cube is of course 1 x 1 x 1 = 1. The difference between the volumes is 0.476401224 (1 – 0.523598776 = 0.476401224). The sphere represents the limit of human knowledge. The cube represents infinite wisdom. We should be thankful that we are allowed to gain more than half (52%) wisdom while less than half (48%) remains unknown. We must always strive to gain wisdom even though we may never gain 100% of it during our lifetime. Human beings are finite in nature while the universe is infinite. Perhaps we may break the sphere and enter the cube after our death. Human life then becomes our half-life. Life after death remains unknown. This is my personal philosophy.

REFERENCES

(1) The Illustrated Encyclopedia Of The universe- ISBN 0-8230-2512-8. 2001 The Foundry Creative Media Company Ltd. London, England.

(2) On the Shoulders of Giants. The Great Works of Physics and Astronomy. Edited with Commentary by Stephen Hawking. Running Press Book Publishers, Philadelphia PA. 19103-4371. ISBN-13:978-0-7624-1698-1

(3) The principia- Sir Isaac Newton, Translated by Andrew Motte, Great Minds Series, Prometheus Books New York, USA- 1995 ISBN-13: 978-0-87975-980-3

(4) Opticks- Sir Isaac Newton, Dover Publications Inc., New York, USA. Based on the fourth edition London 1730. ISBN- 0-486-60205-2.

(5) General Chemistry, Linus Pauling, Dover Publications Inc., New York, USA. ISBN 0-486-65622-5.

(6) Mathematics: The Loss of Certainty, Morris Kline, Fall River Press, New York, USA. ISBN 978-1-4351-3606-9

(7) Mathematics 1001: Absolutely Everything That Matters in Mathematics in 1001 Bite-Sized explanations, Dr. Richard Elwes. ISBN-13: 978-1-55407-719-9, ISBN-10: 1-55407-719-2

(8) Relativity. The Special and the General Theory by Albert Einstein. Authorized Translation by Robert W. Lawson. Three Rivers Press, New York. ISBN 0-517-88441-0.

(9) Discovery of the Grand Unified Theory, Arkay Nair, ISBN 978-1-4634-8731-7 (sc).,AuthorHouse 08/16/2011